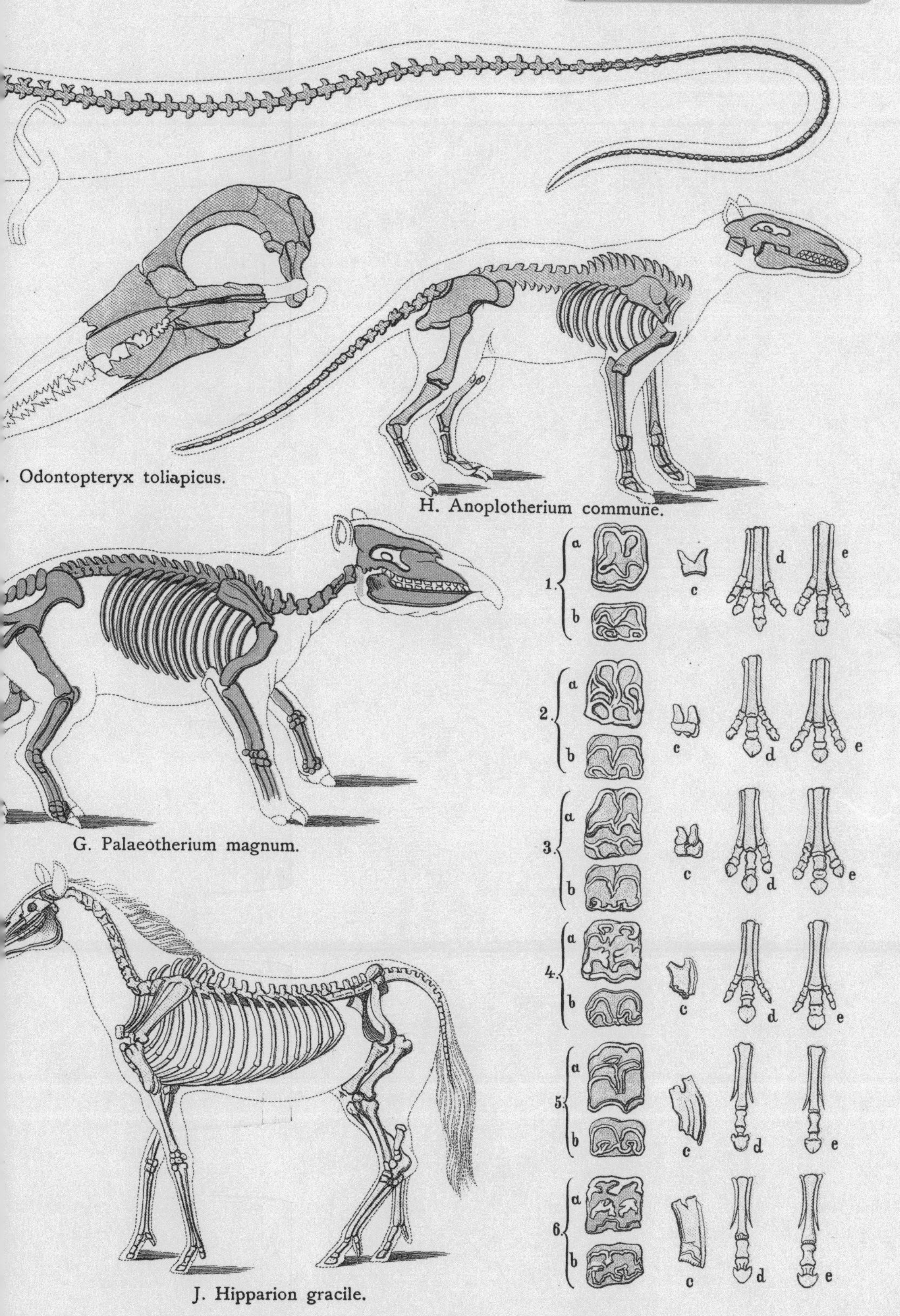

. Odontopteryx toliapicus.

H. Anoplotherium commune.

G. Palaeotherium magnum.

J. Hipparion gracile.

The Call of Distant Mammoths

PETER D. WARD

THE CALL OF DISTANT MAMMOTHS

Why the Ice Age Mammals Disappeared

COPERNICUS
AN IMPRINT OF SPRINGER-VERLAG

Published in the United States by Copernicus,
an imprint of Springer-Verlag New York, Inc.

Copernicus
Springer-Verlag New York, Inc.
175 Fifth Avenue
New York, NY 10010

Library of Congress Cataloging-in-Publication Data
Ward, Peter Douglas, 1949–
The call of distant mammoths : why the ice age mammals disappeared / Peter Douglas Ward.
p. cm.
Includes bibliographical references and index.
ISBN 0-387-94915-1 (hardcover : alk. paper)
1. Mastodon. 2. Extinction (Biology) I. Title.
QE882.P8W37 1997
569′.67—DC21 96-48690

Manufactured in the United States of America.
Printed on acid-free paper.
Designed by Irmgard Lochner.
Cover illustration by Alexis Rockman.

9 8 7 6 5 4 3 2

ISBN 0-387-94915-1 SPIN 10662155

For Chris, Nicholas, and the new one.

Acknowledgments

I would like to thank many people who helped in the writing and production of this book. Three stand out: Jerry Lyons and Bill Frucht, who showed me what really good editors can and should do; and Don Grayson, scientist and scholar. Grayson has defined the debate about overkill and given it meaning and substance. I would also like to thank the excellent production staff at Springer-Verlag, especially Vicky Evarretta.

I apologize for the errors which probably will be found and hope that the imperfections of the writer are no greater than those of the fossil record.

Peter D. Ward
Department of Geological Sciences
University of Washington
Seattle, Washington

Contents

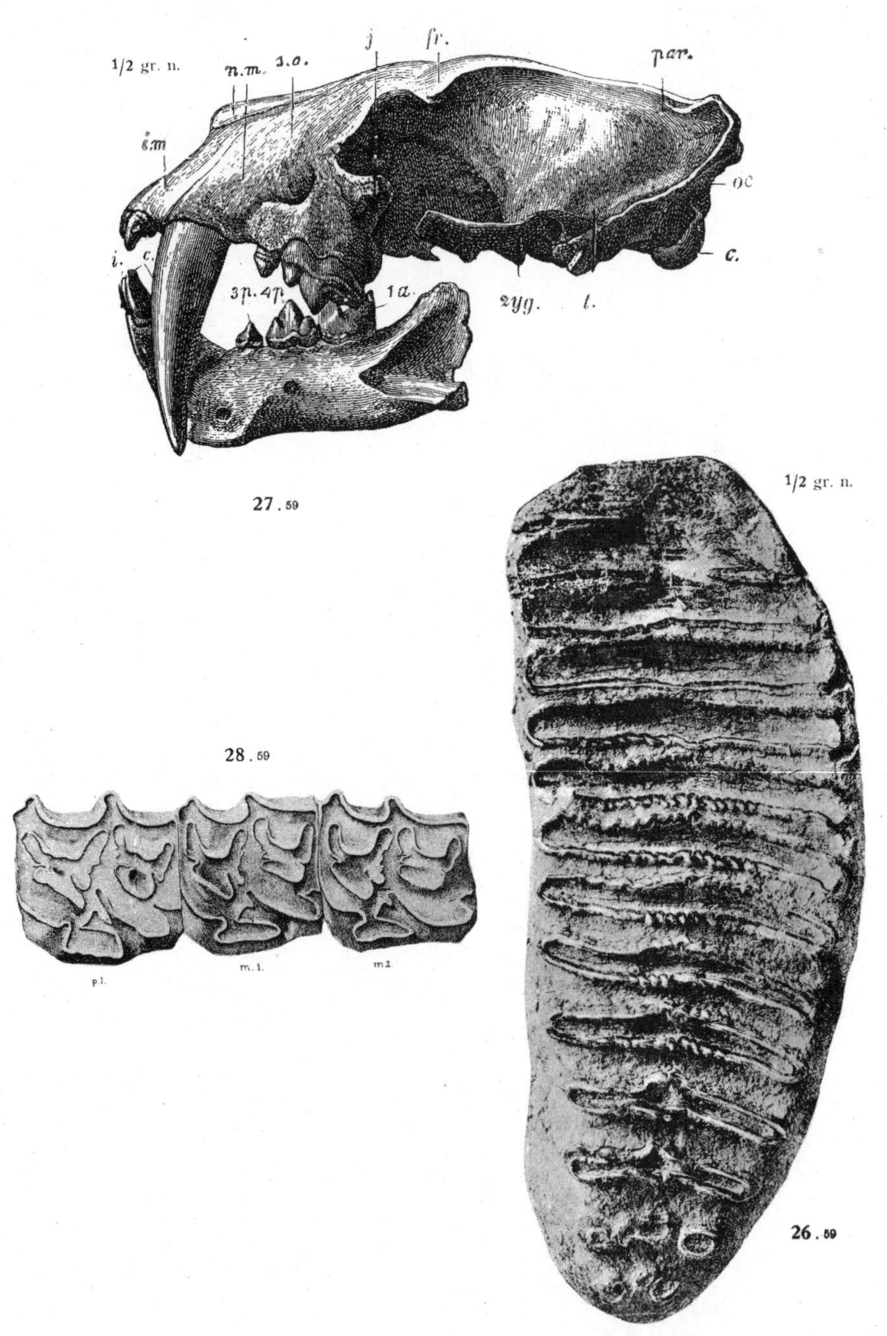

Imp. Tortellier et Cie, Arcueil (Seine)

SKULL OF SABER-TOOTHED TIGER, TOOTH OF MAMMOTH AND HORSE: ALL ICE-AGE "MEGAMAMMALS."

Prologue: The Elephants' Graveyard

There is a tale told to our children about a place deep in the jungle, where elephants go when they are ready to die. This place is called the Elephant's Graveyard, and the story makes death seem dignified and peaceful. A gentle fairy tale, but is there a grain of truth involved?

Elephant graveyards do exist. However, the real ones are not romantic, nor are the deaths occurring there in any way gentle. Some of these graveyards are in the African and Asian jungles, or in the East and South African dry lands where the last elephants live wild today, and where elephants do die, sometimes in large numbers, due to drought, starvation, or human poaching and predation.

Yet, these killing fields in Africa and Asia are not the only elephant graveyards. Another kind exists in the *geological* record—the ancient rocky strata of past ages filled with fossils. These *fossil* elephant graveyards have been found the world over, telling us that once, not so long ago, every continent save Australia and Antarctica was the home of great elephants, and teaching us, as well, stark lessons about extinction.

Finally, there is a third sort of elephant's graveyard, found not in Nature present or past, but in the many Natural History Museums gracing this planet, such as the Burke Museum in Seattle where I work. The dead elephants are brought for study and for safety to these museums. These elephant graveyards are perhaps closest in spirit to the children's tale, for here the bones of both the newly and long-dead elephants are indeed cared for and revered. I have visited such an Elephant's Graveyard each day, to be necessarily surrounded by the bones of great elephants from the deep past. It has always been moving, but some days have been far more poignant than others.

. . . .

My steps echoed hugely as I descended into the stark stairwell, and not for the first time I reflected that a carpet would do wonders here. But this is not a flagship museum, neither Smithsonian nor American Museum of Natural History; it is a typical University Museum, operating on a starvation budget, run by part-time Academic Curators and volunteer labor. The ambiance is less than charming, if nevertheless familiar and utilitarian: here and there one can see a scurrying dermestid beetle (a verminous escapee from the Zoology Department's efforts to strip flesh from bones), while the air yields equal parts of ancient fossil dust and formaldehyde. I reached my floor, deep in the windowless basement of the building, and passed through the featureless halls with fluorescent tubes overhead. I saw our Curator of Paleobotany, in a nearby room, moving rock-filled boxes, and farther down the hall our usually nocturnal Curator of Arachnids sat in his arthropodan lair, surrounded by thousands of spiders, alive and dead. My own mission this day was somewhat different from my normal routine, for I had just received a large fossil tooth, a piece of beach wrack coming from the towering gravel cliffs near the Olympic Mountains of Washington State. It is a relict of the dead, exhumed from its 11,000 year-old grave site by the wind and waves of Puget Sound, and found by an aging beach comber. It has now come to me, and it is my duty to bury such fossils once again, to give them a new resting place, this time not in stony soil but in a tall, gray sarcophagus built by the Lane Scientific Company. Instead of a headstone, these rocks from the Ice Age are given a number, and some vague sort of immortality as electronic life on a large database. It had come to me as so many of these lithic fragments do, not from an organized paleontological dig but from another phone call.

Arriving in my cavernous room, I switched on the lights overhead, revealing the great boneyard around me. This particular room is the final resting place for all of the great skeletons discovered over the decades by both amateur and professional paleontologists of Washington State. The harsh light etches vertebrae and ribs, leering skulls and horns from the near and distant past. But the most striking and spectacular objects in the room are the great ivory tusks—curving, giant elephant-like tusks, far larger than any elephant tusk of today and in unlike shapes—for this room is filled with the remains of mammoths and mastodons from the Ice Age. It is to this graveyard that I have brought a new piece of the old. Starting the slow process of curation, I replayed in my mind the events just transpired two floors above.

The telephone call had come the previous day. Being one of the few paleontologists in a state exacts its own unique price to pay, for to the unnumbered children in my state, every round rock is a hopeful dinosaur egg, every old decaying cow bone the humerus of an *Allosaurus*, every stony shell the promise of an ancient world; yet sometimes, just sometimes, these extravagant promises turn out to be correct. Thus, I never refuse the phone calls about some newly dreamed dragon, for the rare dragon bones do exist, to be disinterred each spring by the steady Northwest rain.

So I had answered the phone, to receive a most unusual request. A woman caller hoped that I could see her father. He was terminally ill with cancer and in his eighties. Furthermore, could I see him . . . soon. I asked why. Her father, it turned out, had discovered a fossil in the gravel cliffs of the Olympic Peninsula many years ago, and for the last decade had become fixated on it. He had pored over numerous books, and had come to believe that it represented a new species of sea creature from the Paleozoic Era of more than a quarter billion years ago. I told her that I would be happy to see him in the following week. The woman paused, and then asked if I would see him . . . the next day. She was not sure that he had a week of life left. She also asked if the entire family could attend, as her father wanted to give his fossil to my museum. Many words passed between us in the silence that followed, and I answered that I would be delighted to see him.

They arrived the next morning, in a small caravan of cars; three generations of a family brought together to see the ancient man on his life's last outing. They had also called the press, I saw with dismay, to watch the old man turn over his ancient prize. We met in front of my museum's giant Ground Sloth skeleton, itself a wondrous relict of the recent Ice Age of 11,000 years ago.

The man in his wheelchair was frail, skeletal, and lined with age; his skin a yellow parchment of history. Yet I knew that only short years before, this same man had been young, with dreams of life and the presumed immortality of youth cloaked about him. But in seventy to eighty years, life had passed through this man, and death was very near. Not for the first time I cursed the gods for making us merely mortal, for damning us, alone among animals, with certain knowledge that this greater gift of life and consciousness is only lent, not given.

He held a large box in his trembling hands, and I yearned to see inside, to once more use the past as an escape from a painful present, for life now clung to this man by only tenuous threads, and I, student of death that I am, hypocritically quailed before the Master of my profession. He wanted to see the other fossils of my museum first, and I wheeled him from exhibit to exhibit, telling stories bringing back the past. We finally stopped with a sigh from him. He began to explain his theory to me, how his find, made twenty years before, had sent him on a great quest of intellectual discovery, spanning many libraries and sources, and how he came to realize that he had found a new type of Bryozoan, a filter-feeding creature from the sea not uncommon now, but once ubiquitous on the sea floors of Paleozoic days. I saw him now as another time traveler, another who has journeyed back to remote prehistoric days; and I felt a great bond. Bill Dietrich, the Pulitzer Prize-winning science reporter from our local paper, was near us, quietly waiting to see this new find. The man opened the box and pulled out a misshapen lump of rock from within. Bill looked at me, and I gave him a quick head shake "no," and he quietly packed up his camera. The rock was half the size of a bowling ball, and falling to bits; it was layered and was clearly once part of the living. I asked the old man to describe to me exactly where he had found it, and he told me that it had eroded out of gravel cliffs near Port Townsend, a small town hugging the foothills of the Olympic Mountains. He handed me his treasure, touching it for the last time, and I could see where it was worn from countless hours of his tactile examinations. He finally asked me if he was correct in his identification, and I debated for a moment; yet there is truth and truth. I told him that he was absolutely correct in his interpretation; and for a few moments life seemed to fill back into him, enlarging him. He gave a smile of triumph to his family, and was slowly wheeled away. I never saw him again. As the family gave their quiet thanks and filtered out, I repacked the fossil and carried it out of the public area, to its waiting grave below.

Alone now, amid the brooding skeletons of earth's ages past in my museum's prehistoric elephant's graveyard, I looked once again at this newly proffered fossil, and felt the chill that comes from being in the presence of the ancient dead, and from being in the presence of wonders. This fossil was unrelated to any sea creature; it had never spent a day beneath the sea. Yet, to me, it was something far more powerful than the old man had even realized. It told of a recent time when elephants roamed all the earth, not just Africa and a vanishing part of Asia; a time of great ice sheets strangling the land while our ancestors struggled to keep our species alive; and briefly, not for the first time, I imagined I heard a distant call of mammoths. It was a tooth of a mammoth, perhaps one of the last of those great elephants to have lived in North America, some 11,000 years ago, a faint echo coming from the time when virtually every human on earth knew the familiar call of distant mammoths. It was also a symbol of how powerful the pull of the past is, and how important it is to understanding our present and future.

Any piece of the past is a wedge into the our planet's history, a slice of the long ago; any fossil is one more small cog that, when meshed with many other paleontological pieces, builds what we might call a metaphorical time machine, giving us a peek into time long gone by. In this case, the Time Machine has yielded a view of the world that had yielded this ancient molar, a world quite unlike the environment of modern day Seattle or Vancouver. The lithic tooth I had just received surely once chewed the green bracken and shrubs lining the newly emerging Puget Lowland; it was part of a Columbian mammoth, or perhaps even a Woolly mammoth, which had lived along the western shore of the then, newly created inland sea now known as Puget Sound. The high Olympic Mountains were at that time still sheathed in glacial ice, and the sea itself was dotted with stagnant ice sheets and freely floating icebergs, but the great, mile-thick sheet of continental glacier that had covered half of North America and all of the Pacific Northwest north of Tacoma some 15,000 years ago, had, by the time of this mammoth's life, retreated northward, or into high mountain keeps. This tooth came from an animal living in a land newly emerged from its ice cover; from a warming land, from a paradise for big game—and for the predators, both two- and four-legged—who stalked the big game.

This mammoth was surely one of a large herd, its kind common here among the New Growth forests of Cedar, Douglas Fir, and scrubby alder trees. The mammoths were not the only large mammals in this region; giant ground sloths, almost reptilian in appearance, were common as well, as were deer, elk, moose, bison, and the great predators and scavengers that lived on the huge herbivores. It was a land and time of change, the transition from Ice Age to a new age, a time so new as to have no name other than "the post glacial." It was also the last halcyon moments before a great period of extinction, for most of the great mammals of 11,000 years ago were to be the last of their kind. Eleven thousand years ago, all over North and South America, they thrived. By 10,000 years ago, all but a relict few were gone. Their passing remains one of greatest paleontological mysteries of all.

Once more I look at the lump of tooth in my hand, so well worn first by the elements, and then by the hands of a man. But I wonder, too, if the old man's hands were really the first human hands to have touched this tooth; I wonder if this fossil, and the great elephant which produced it, were not wrenched from the living and sent to the dead by another hand of Man, some 11,000 years ago. Did this particular mammoth die from some "natural" cause, such as drought, or lack of food, or disease; or did a murder end this particular life? Was this particular mammoth's last view of this earth not of its children, or members of its herd, but of two-legged hunters, armed with spears, brandishing fire and screaming in triumph as triangular chert spearpoints sought the great pulsing heart; did this mammoth have a last moment of life watching its life-giving liver being torn from its body, and given in bloody triumph to exultant, hungry human hunters? That is the greatest mystery: did *our* species consign the mammoth, and so much else, to a long-ago elephants' graveyard? And can it, *is it* happening again, to other elephant species, in the present day, and in the near future? The answer to these riddles, one in the past, one in the future, can be found only with a time machine—the paleontological time machine of research.

The premise of this book is simple: very recently in the geological past, a great catastrophe took place. That catastrophe, among other things, caused a great number of large mammals and birds to go extinct in very short order. I believe that this great catastrophe, which can also be defined as a mass extinction, removed from our Earth all of the mammoths, mastodons, ground sloths and saber-toothed tigers, among many other storied creatures, and that this episode that has so robbed the earth of breathtaking creatures *cannot* be understood as extinction through "natural causes," *cannot* be explained by slow climate change, *is not* just one more gradual change of evolution. It was a catastrophe aided, or even largely caused by the influence of humankind. It was probably savage and fast. It can best be understood by viewing other catastrophes, other swift executions produced by rapid catastrophes occurring long before the evolution of humanity. One such catastrophe wiped out the dinosaurs 65 million years ago. Another may kill all surviving elephants over the next millennium. The present is *not* the key to the past, but may be the poisoning of the future.

As long as I have been a professional paleontologist I have believed that to understand the past, you must first know the present; that a firm understanding of the biology of extant creatures is a necessary prerequisite for interpreting the lives of their now-extinct relatives. Paraphrased, any scientific time machine that lets you see into the past—be it a rock pick on an excavation, a modern radiocarbon dating lab, or a powerful computer analysis, among others—uses the *present* to calibrate and study the past; the time machine always starts in the here and now, and then travels *back* through time. But for the first time I am awakening to a new insight—that studying the *past* may hold the key to understanding not only the present but the future as well. Perhaps, the time machine can, and must move *forward* in time, as well as back into the past. Perhaps it is the past, only the past, that can teach us lessons necessary to avoid a catastrophic extinction of the last elephants, and much else, in the near future.

Paleontology is an inexact science, filled with uncertainty. But of one thing I am sure. Catastrophe is surely the key opening the gates to the elephants' graveyards.

The Call of Distant Mammoths

Why the Ice Age Mammals Disappeared

Mountains and glaciers. (From L. Figuier, La Terre Avant Le Déluge, 1864.)

1

The Time Machine

RAIN DEFINES THE NORTHWEST. Sometimes it comes as a heavy drumming downpour, with a cadence almost Brazilian; sometimes it brings a rare rainbow. Usually it falls as a gray mist or as shrouded sheets billowing downward, coating North America's Pacific Northwest coast with a dark sheen. Sunny days, of course, do occur, often enough to nourish plants and animals and the human spirit as well, and even snow may whiten the land on the occasional winter day, but these variations are spacers between the rainy periods. The rain is gentle, often warm, yet insistent; it is the true chronometer of this land.

The rain falls mostly on a land of geological youth, a land of gravel. The high mountains are exceptions, bastions of the old. Even the comparatively young parts of the central Cascade Mountains, the volcanic rocks that make up Mount Baker, Mount Stuart, and the dangerous twins Mount Rainier and Mount St. Helens, have existed for millions of years. But in the North Cascades, ancient rocks left behind from the Age of Dinosaurs struggle skyward, and farther north, in British Columbia, quarter-billion-year-old assemblages constitute the skeletal backbone of this part of the continent. Like the few remaining Douglas fir and western cedar that tower above the much younger second- and third-growth broadleaf forests, the great rocky mountains looming over the expanse of gravel that covers so much of the Pacific Northwest are rare relics in a recently transformed landscape.

The rain falls on these old high Cascade Mountains and then starts its rush to the sea. This new phase begins in boulder-strewn rivers, places choked not only with rocks disassembled from the nearby mountains but with alien rocks as well. The river beds contain mineral types clearly incongruous here, such as the granite, schist, and gneiss that appear in places where no mother lodes of like rock exist, and the great rounded blocks of sandstone packed with fossil shells found nowhere else in the region. These alien rocks are evidence of the recent Ice Age. All of the strange, loose rocks found today in the Pacific Northwest were carried here from the north in a sea of moving ice that more than twelve millennia ago scoured the land.

It is the fate of these rocks to crumble and move, for all eroding rocks, native and migrant alike, are sooner or later taken up by the high mountain streams as they carve their way to the sea. The rocks are daily breaking down, weathering, moving slowly at first but accelerating over the years as they crack and fragment into ever smaller sizes, eventually adding new gravel to a land already covered with it.

The rivers rush downward and westward at first, carrying their cargo of rolling rocks, but they soon slow down, widen, and increasingly meander as they reach the great inland valleys of western Washington and British Columbia. Here they grow, through the intersection of fast streams like the

White River and Icicle Creek, into the broader and more placid Cedar, Green, and Frasier Rivers, flows that a century ago snaked across sprawling lawns of lowland old-growth temperate rain forests. The landscape we see today is largely manufactured, but even after a century of clear-cutting and human development, one constant remains: Everything—giant cedars, lush pastures, shopping malls, and housing tracts—rests on beds of gravel, boulders, rocky cobbles, and poorly mixed sand grains. If the trees, soil, and human artifacts could be stripped from the Puget Lowland, there would remain only these fields of disorganized gravels, not bedrock; the true rocky foundation of this land lies hundreds or thousands of feet under the vast glacial till that makes up the Northwest. It is a geologist's nightmare, for all this landscape tells us is that once, not so long ago in geological time, glaciers advanced and retreated, leaving only chaotic gravel behind.

Much further in the past, the regions now covered by the spreading cement of Seattle, Tacoma, and Vancouver held great inland seas and river-covered plains ruled by creatures far more ancient than any found there today. Once the rivers of this region were lined by huge forests of palms, not cedars, and five-toed horses the size of squirrels foraged among the trunks of these tropical trees or fled the carnivores of the Eocene Age. Fifty million years ago, the nearby sea bathed seashores in tropical warmth and harbored not the orcas and salmon of today but whales with distinct necks that held their heads up high. The heads of these carnivorous whales brandished huge, triangular-pointed teeth. On this deep Eocene sea bottom, snails and clams in carnivals of color and shape proclaimed the diversity of the tropics. It was a place that would not appear entirely unfamiliar to a Jamaican or Fijian of our world but would surely seem alien to any modern-day Seattleite. It was a land where rocks existed as stony outcrops, not gravel piles—rocks whose age would have been properly measured in millions of years, not the paltry thousands of years that tell the age of the gravels that blanket the western half of Washington state today.

Once this region was indeed like any other, but then the glaciers came, covering the landscape first with ice and then, as they finally loosed their grip on the land, with gravelly residues. The glaciers scoured or buried the older landscape as if by whim, slicing huge rents in the original bedrock and depositing great foreign boulders from places farther north or carrying local rocks southward. The great glaciers were bleak angels of death and cold. Enormous black rivers of ice crawled south from Canada or crept down from the high mountains to join with the more sluggish lowland glaciers in frigid embrace, a slow-motion duet crushing the land, burying forests, driving away the rich game. As recently as 15,000 years ago, ice a mile deep entombed the future home of Starbucks and grunge bands and Bill Gates. Now, these many years later, the only legacy of that time is a land covered by gravel and a countryside with very few larger mammals left.

This gravel hides secrets of the Ice Age, and most of these secrets are of long-ago life and death, of ancient arrivals and ancient extinctions. Only a time machine can bring those days back, and often the Northwest rain serves as that time machine. Relentless centuries of rain have sluiced away the piles of boulders and slabs and pebbles and cobbles; and occasionally treasures emerge, telling us of the land of ice that was so recently here and of the great drama that unfolded when the ice sheets finally melted away the last time, only 12,000 years ago. New inhabitants came from the far north along the newly ice-free corridors, invaders from Siberia via Alaska searching for more food and a better life than that afforded by the bitter winter of their northern homes. A 2000-mile-long valley emerged just east of the Canadian Rocky Mountains as the glaciers melted way, linking the Northwest with the subcontinent called Beringia, now submerged beneath the Bering Sea. Beringia, the ancient land bridge connecting Asia and North America, was home to skillful two-legged hunters armed with cruel spears. These people had originally come from Asia, where they hunted the great woolly mammoth among the other game, and as they moved ever southward in a landscape with little vegetation, they surely became some of the most efficient hunters ever to have lived on this planet.

They arrived in the Pacific Northwest less than 12,000 years ago and must have been delighted by the great herds of beasts that they found as they entered the rain-drenched land. Their journey had been long and arduous; they surely arrived desperate, with children they loved and precious mouths to feed, perhaps with dreams of shorter, less cruel winters and more abundant game. In this hope they could not have been disappointed, for they encountered a countryside redolent of flowers and teeming with game; they found a warming, ice-free land rich with creatures that had never seen humans.

In the 2000 years following their arrival, these people, whom we now call the Clovis, became the ancestors of all of today's native North and South Americans. In those two millennia they reached the farthest corners of both continents, spreading even to the southernmost tip of South America, and with fire and hunting forever changed the landscape. In the same short time, another, even greater change took place: Over 50 species of large North American land animals, and perhaps even more in South America, went extinct. To many scientists it is no coincidence that the coming of the Clovis coincided with this enormous biotic crisis. This was something new in the history of humanity. Our species had reached new areas many times before, and local extinctions had indeed occurred after our arrival, but none so suddenly or so catastrophically. And therein lies a paleontological mystery: Were the Clovis the most efficient hunters in our species' long history and the greatest single cause in the most catastrophic large-animal extinction since the death of the dinosaurs? Or was their arrival a minor element in a mass extinction already under way?

The Northwest rain surely fell on the Clovis, a rain that swept away the blood of the freshly killed game and perhaps muffled shouts of joy as a hungry people brought down their first North American mammoth, an animal at first unaware, and perhaps too long unafraid, of the rag-tag band of bipedal creatures newly arrived in this green land of gentle temperatures. The rain covered hunters and hunted alike and nourished the small streams that eventually pushed glacial gravel over the butchered elephant's corpse, to bury it for twelve millennia.

Finally, the rain resurrects these long dead. Each spring, the gray Northwest rain sluices aside the Ice Age gravels to exhume a new cache of ancient elephant bones, each more precious than any gold nugget to the few academic time travelers in the region; the constant rain brings forth ivory pearls of age and wonder to tell us stories that refuse to stay buried—the stories of the ancient elephants' graveyards.

. . . .

As recently as 12,000 years ago, much of the world's land areas were inhabited by a bestiary that might have been mistaken for the set of a Tarzan film: Our world had a distinctly African flavor. Then, over a period perhaps as short as a single millennium or as long as five, a great extinction took place among the larger birds, reptiles, and—above all—the large land mammals on all continents save Africa. The few larger animals found today outside Africa are but a shriveled vestige of the Ice Age megafauna. We humans missed seeing a wondrous biodiversity by the briefest instant of geological time.

Imagine that you awoke a millennium or so after the last dinosaurs had perished and that all that was left of this great saurian history was whitened, bleaching bones eroding from gravels and stream banks, sand dunes, and lake beds: bones from animals so recently dead that they can hardly be counted as fossils. If you were a ten-year-old, you would have been extremely annoyed at having just missed the Mesozoic icons; even adults might ask for their money back. Yet our situation today, in the twentieth century since the time of Caesar Augustus and the fortieth since the last mastodon perished on a remote island near the Arctic Circle, is similar. We ourselves have just missed a large, equally impressive fauna far more diverse than any assemblage of dinosaurs from the Mesozoic Era.

What happened to this great Ice Age bestiary? Why are the mammoths and mastodons, giant ground sloths and saber tooths, woolly rhinos, cave bears, Irish elk, elephant birds, moas, giant lemurs, enormous terrestrial crocodiles, and giant kangaroos now known only from dusty exhibits in natural history museums? Unlike the dinosaurs, which had the bad luck to inhabit a planet that was on a collision course with an enormous comet, the Ice Age megafauna died neither simultaneously nor from any universally acknowledged single cause. There are currently two competing theories to explain this most recent of mass extinctions: Either the climate changed drastically at the end of the Ice Age, killing off the great beasts in the process (the Climate Change Hypothesis), or they died of what we might euphemistically term "unnatural causes" and any detective would call "foul play." This second theory, that Stone Age people directly caused the single greatest mass extinction of large animals in the history of the earth, is called the Overkill Hypothesis.

The view of this past is nowhere better displayed than in downtown Los Angeles, California. There, in the La Brea Tar Pits, a great deposit of naturally occurring tar and oil, hundreds of thousands of bones from the Ice Age have been excavated. There had been no glaciers in Los Angeles for hundreds of millions of years before this, no sheets of ice or sediment-choked rivers, no tundra or steppe or any other hallmark we associate with the great Ice Age. Yet there were great beasts in plenitude: saber-toothed tigers and wild horses, giant ground sloths and camels, hippos and lions and enormous scavenging condors, great bears and giant wolves—a vast diversity of a kind today associated only with Africa. Most splendid of all, the urban metropolis of Los Angeles was then populated by great herds of giant, now-extinct mastodons and mammoths. The area was thus home to some of the largest land mammals the world has ever seen, the biggest animals to walk the earth since the time of the dinosaurs.

Near the end of the Ice Age, the world was rampant with proboscideans, the elephant-like mammals. The familiar African elephant, whose scientific name is *Loxodsonta*, lived then as now only in Africa, and Asia was home to great herds of Indian elephants, or *Elephas*, as well as to mammoths and mastodons. North America had both mammoths and mastodons, whereas South America was home to three genera of proboscideans, all about the size of modern-day elephants, of a type called gomphotheres. Gomphotheres had both upper and lower tusks, unlike elephants, mastodons, and mammoths, which had tusks on their upper jaws only.

There is a common notion that during the Ice Age, the world was everywhere locked in great sheets of ice. My first image is of mile-high glaciers and frozen steppes; of great hairy mammoths, mastodons, and rhinos trudging dourly into frigid Arctic winds. But the reality for most of the earth

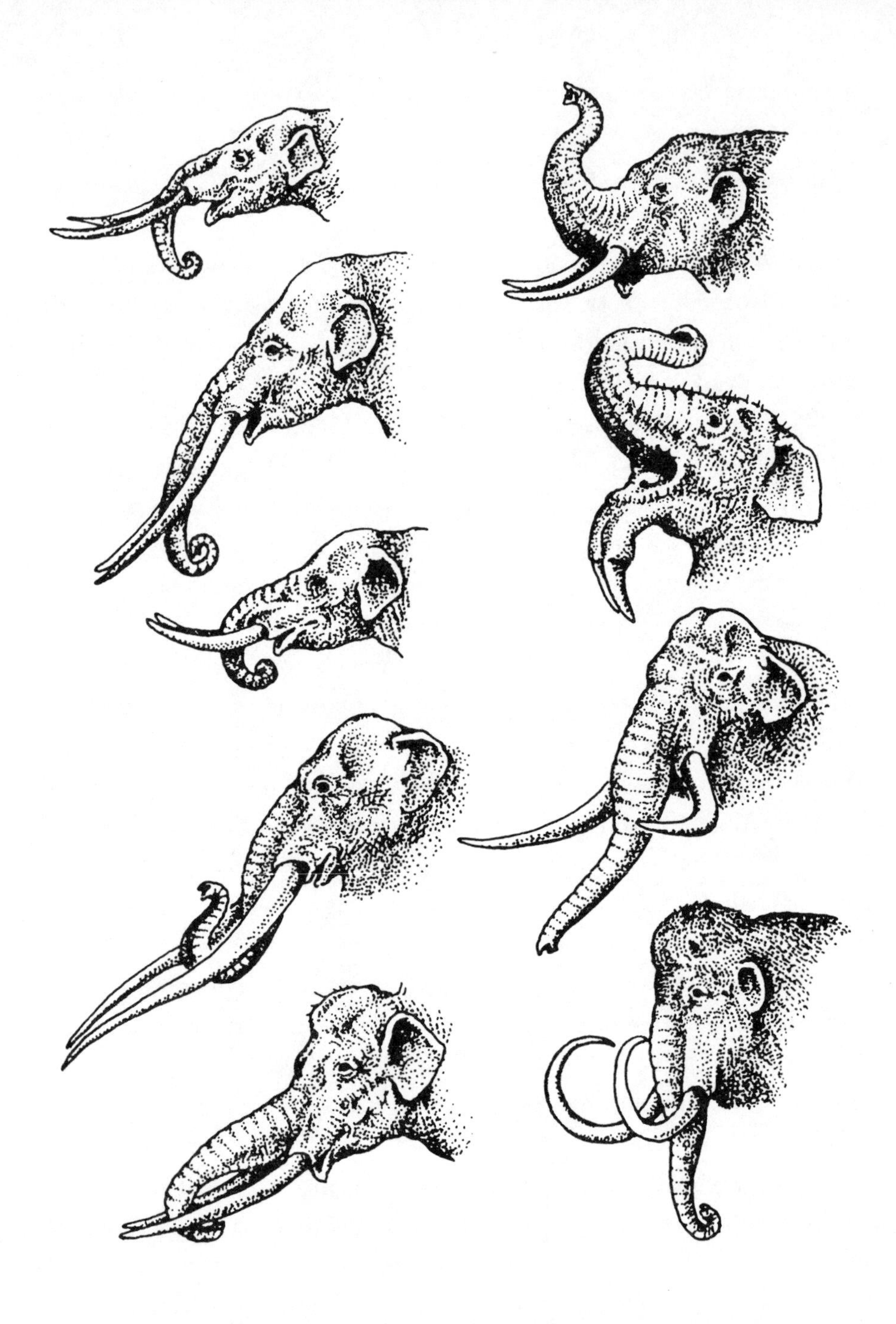

Ice Age elephants. Extinct Pleistocene Proboscideans and Deinotheres. Left, top to bottom: *Cuvieronius*, *Stegomastodon*, *Haplomastodon*, *Anancus*, and *Stegodon*; Right, top to bottom: *Mammut*, *Deinotherium*, *Mammuthus* (*Archidiskodon*) *meridionalis* and *Mammuthus primigenius*. (From Martin and Klein, eds. Quaternary Extinctions, 1989; with permission.)

was very different: There were plenty of spots perfect for a Club Med or for a good snorkel on a tropical reef; there were deserts and rain forests and most environments that we find on earth today. Take Los Angeles. If you could visit the Los Angeles basin of 12,000 years ago, you would still need good beach attire; you could still get a great tan at Malibu (never mind that the sea was more than 100 feet lower than it is today, so the beach would be a bit west of its present position); you could hike up to the San Gabriel Mountains in bright sunshine. But you would be wise to bring a big rifle along, or better yet, a few hand grenades. Saber-toothed tigers were plentiful, and there might be the rogue mammoth to deal with. . . . But we are getting ahead of ourselves. First we need a time machine for the journey.

. . . .

If you were to build a time machine, what would it look like?

The time machine story, and time travel in general, are a rich tradition in the genre of science fiction. For my money (and I have spent my share on the subject), the best machine of all was the wonderful, ornate creation of H. G. Wells in his great turn-of-the-century tale called, directly enough, *The Time Machine*. George Pal's 1960 movie version did full justice both to the machine and to Wells' wonderful cautionary fable. (Who could top Rod Taylor's world-weary look as he begins to realize that the future of humanity may be no better than its past?)

Innumerable follow-ups on the theme have since emerged in print and on the screen. Physicists even speculate today that time travel (of a sort) is theoretically possible; around the edges of black holes (according to calculations that most of us have to take on faith), time can run backwards for the briefest of instants. But such ideas are of no comfort to me; I need to travel far longer than a brief instant backwards in time to see what I want to see. A lifetime wouldn't be enough, nor would a paltry millennium.

There would be so much to learn: Was *T. rex* a scavenger or a predator? Could *Archaeopteryx*, the first bird, fly, or was it a feathered ground runner? Did ammonites have four gills or two? Did sauropods live in groups? Did hadrosaurs sit on their nests? The list is infinite. Where would we start?

Clearly, we would have to make a lot of choices. First among them would be how far to retreat in time. Would we visit some time in the historical past of *Homo sapiens*, our species, or go back farther, to a time when we existed but had no written language to record how grand (or how awful) life really was? Would we go back to a time when our species was still part of the "natural" world, before technology made us the big winners in the game we call evolution, where every species tries to carve out as big a piece of the earth's available resources as possible—a game now won hands down by *Homo sapiens*. Or would we go back farther yet, before the first appearance of our species, to the great, late summer of the Cenozoic world, before the climate cooled and giant Pleistocene glaciers reconfigured the planet of the Pliocene Epoch? Would we plumb the depths of the Cenozoic Era, when horses had five toes and our squirrel-sized ancestors lived in trees? Would we go deeper yet, back into the Mesozoic Era, to the birth of flowering plants and the heyday of the dinosuars, to test Steven Spielberg's computer-animated conceptions of the dinosaur fraternity? Or would we insist on seeing more archaic vertebrates, back in the time of splayed legs and dragging fish-like tails, or back past swampy coal ages into clear coral reefs dominated by squirming, inefficient piscine predators and marauding spiny arthropods? Would we be like Captain Picard, in the last episode of *Star Trek: The Next Generation*, who went with Q to see the primordial soup in which Earth's first DNA molecules—the first life—began some four billion years ago? Or would we travel even farther back, to the early formation of Earth, when a Mars-sized world collided with our nascent planet, ripping us asunder and in the process creating our moon? Or back farther still, to the time of the formation of the Universe, to watch God light the fuse of the big bang? Indeed, there are many choices.

But of course this is nonsense. There are no time machines other than our minds, our rock picks, our libraries; the only modern time travelers are the paleontologists and archaeologists who succumb to the temptations of ancient grave robbing, of disturbing the long dead, of prying the cold bony fingers of time from around the treasure chest of history; of wresting fossils from their stratal graves. And all our speculations are dreams of ancient evenings, which often evaporate in the harsh morning light of objective scrutiny. Proof, the most hard-won prize in science, is even more elusive in paleontology and archaeology than in other branches of science. Sometimes the best we can manage is intuition.

How much about the world, beyond the obvious differences in human technology and population size, would be different if we went back a only a thousand years? Would we notice any difference in the color of things, the shapes of clouds, the bugs underfoot? Would the composition of trees, the types of flowers, or the songs of birds tell us we had entered a different world? I doubt it.

What about farther back, beyond a puny millennium? How about well into the past, into the Mesozoic, the time of the antediluvian kings of creation and box office, the dinosuars? Here, surely we would notice enormous differences. In the Mesozoic Era, starting 250 million and ending 65 million years ago, the very atmosphere was different, with higher amounts of both oxygen and carbon dioxide. The seas were different in terms of their variation in oxygen content and temperature from top to bottom, as well as in the nature and pattern of the ocean's circulation. These two factors might give the atmosphere a different look, or color; perhaps the clouds themselves would seem different to us. And surely we would notice the complete lack of flowers, fruit trees, grass, large mammals, and abundant birds if we were as far back as the Jurassic Period.

If we went even farther back, into the Paleozoic days of more than a quarter-billion years ago, we would encounter stunning differences. The rotational rate of the earth was radically different. The moon was closer. Few modern mountain ranges had yet emerged. The seas were mainly shallow and warm, fetid basins. The largest land creatures roved on six legs or more. The oceans were filled with creatures of unimaginable colors and shapes, spindly forms of our world's nightmares. Mammals, and humankind, were not even a distant dream; surely our time machine would deliver us into a strangeness beyond imagination.

Where do the realms meet? When did *Time* become filled with things and conditions that we would recognize as being of our own world? I have always believed that the world took on its modern form during the Cenozoic Era, the so-called "Age of Mammals" that began 65 million years ago. Now I am not so sure. Perhaps the world we reside in is not older but *younger* than we think. Maybe we are not rooted in deep time, and strangeness awaits us only small steps back, only millimeters beneath the most recently developed zones of our cerebral cortex. Perhaps we need only return to the Ice Age to find a strange new world.

The world has been in an Ice Age for a longer time than any member of our genus has walked this earth and for ten times longer than *Homo sapiens* has existed. Recently discovered evidence suggests that the world immediately prior to the onset of the Ice Age was a very different place from the one we know. Less than a million years ago, hippopotamus were common in the Thames estuary, where the great city of London now stands. Just twelve thousand years ago, mammoths and mastodons walked on much of North America. Perhaps only the most recent of millennia would appear at all familiar to us. Certainly, to visit each of the major continental or large island masses during the latter parts of the Pleistocene Era, the time of the last great Ice Age, would be to experience a sense of disorientation simply from the makeup of the terrestrial fauna. So many huge creatures have disappeared since the end of the Ice Age that we now live in a highly impoverished world. To return to the world before their disappearance might be the most sensational time trip of all.

So I have my time machine revved up and ready to go. It is fabricated from the numerous complex analytical instruments as well as the simple hammers and spades of my trade; it is equal parts paleontological digs and archaeological expeditions, articles and lectures by an army of scientists; it is the product of the scientific method. There is no way of knowing whether the world it takes me to really existed, for paleontology is often a best guess. But I will take that chance. My time machine is set for three times and places: South Africa of 115,000 years ago, Australia of 35,000 years ago, and North America of 11,000 years ago.

. . . .

The southern coast of Africa, showing Table Mountain and Cape Town. (From The Life and Explorations of David Livingstone, LL.D.)

The thing didn't work. It is the only possible explanation. Everything looks so familiar! I have taken the longest voyage into the past, well back into the Pleistocene Epoch 115,000 years ago, according to the dial on my time machine, which was geographically set for the southern tip of Africa, slightly to the east of the Cape of Good Hope. I have arrived. I am clearly in Africa and on a stretch of beach I may have once visited in my own world. (Or is it a place I *will* visit? Time traveling is confusing.) Has the machine transported me? Am I in a reasonable reconstruction of this particular past?

I have landed on a beach with thick, buff-colored sedimentary rocks (Paleozoic Cape Series, I tell myself smugly) cropping out in quite normal fashion. The trees and other vegetation are exotic to me, simply because they come from one of the most distinctive of all floral provinces, the Cape fynbos, a vegetational assemblage composed of many species found nowhere else on earth. It is dominated by *Proteas*, those huge, thistly flowers in a rainbow of colors that always look artificial in a florist's shop; here they cover great bushes and trees. It is a warm, sunny day; the ocean is blue, and hundreds of sea birds whirl overhead. It is very difficult to believe that this is the fabled Ice Age.

The intertidal shoreline is crusted with mussels and barnacles, the beach showing the wrack of the sea: kelp, a jellyfish or two, and hundreds of stranded cuttlebones, the internal skeletons of the squid-like cuttlefish that live in remarkable numbers in the shallows here. Mixed forest and grassland come right down to the beach, a confusion of semitropical trees. Is this my world or the world of 115,000 years ago? How can I tell? I begin to search the sky, looking for jet contrails. None, but how often does that happen in our own world in any 5-minute stretch? I see no roads or other evidence of the modern world, but I know there to be large stretches of national park along this section of the South African coastline. The Otter Trail in the Tsitkara Park is an excellent example, and for all I know I could be in that park right now, just as I was (or will be?) in 1992. But soon a sad reminder creeps in: a good method of detecting the stage of human civilization. I am on a beach, after all, and beaches the world over in the 1990s betray all too accurately the human presence. No beach in our world, no matter how remote, is exempt from the water-borne evidence of humanity. The seas are the dumping ground for ships' refuse—and land-based garbage as well—and the oceans' great circulation patterns have efficiently distributed the material evidence of our existence. Candy wrappers, plastic gags, Clorox bottles, tampons—they all float and never degrade. They have spanned the globe, ensuring that every beach remains well stocked with the stratified flotsam and jetsam of humanity. Even if at first you see no tell-tale human wrack on any given coastline, just dig down into the sand.

An even more perverse thought then enters my head: In early 1996, a report began to circulate on the Internet about the existence of a huge circular mass of condoms in the Pacific Ocean, trapped by currents equivalent to those that lock seaweed in the Sargasso Sea. Supposedly this new island is composed of a mass of latex 20 miles long and 6 miles wide, millions of condoms drawn together by their similar size and flotational properties. This story must be the product of someone's imagination, but it rings true, doesn't it? They don't disintegrate for hundreds of years, and they must go somewhere. In the earlier parts of this century, there were rumors of the "Flying Dutchman," a lost ship carried by the currents. Today, we tell tales of the Island of Floating Condoms.

I begin to peek among the frothy intertidal, looking for relicts as well as relics, not really caring whether I am in the present or the past, enjoying the warm day on a perfect beach, listening to the soothing whisper of the surf intermingled with the calls of monkeys and birds, the remote trumpet of an elephant, and . . . the growl of a nearby cat? Is that a lion? Now wait, the South Africans do have exquisite game parks inland, but I have never heard of a great cat this far south. Come to think of it, my foot-dragging sampling of the strand line has come up empty. Perhaps, this is a well-policed beach, for there is no human pollution here, no evidence whatsoever that Dustin Hoffman has yet heard, as he will in the 1967 movie *The Graduate*, that the future can be summarized in one word: "plastics." How likely is that?

So here I stand on this beach, debating with myself, *maybe it is*, *maybe it isn't*, dressed in comfortable clothes, wearing a good pair of boots, maybe in our world, and maybe back 115,000 years, in a place and time I have chosen to visit for one reason: The fossil record suggests that this is one of a few sites on Earth where the first creatures of our species, whom we call *Homo sapiens*, or the Moderns, are known to have occurred. Fossils of this oldest known human—creatures anatomically indistinguishable from us in all major characteristics except, perhaps, the size of their teeth—have come in significant numbers from Di Kelders Cave, near the Cape of Good Hope, from Nelson Bay Cave, located farther east, and even from inland, from Equus Cave and Florisbad, both in the great Karroo desert, home to the oldest mammals ever recorded. But by far the most significant site is on this fabled southern coast of Africa, at a place called Klasies river. Elsewhere on earth, other varieties of hominids were dominant at this time; over much of the world, late varieties of *Homo erectus* were common, and in a few places in the Middle East and Europe, the Neanderthals held sway. But here in southern Africa, evolution had wrought a much different creature from the latter two, a band that would soon take over the earth and that, in a short 115 millennia, would reach not only the farthest corners of the world but the moon as well. Perhaps here, in South Africa, or perhaps farther north, in the Rift Valley of East Africa, humans were born.

Thus either in our day or back in time, I decide to sashay along the beach to see what there is to see. This being my first time travel, I am uncertain about what to bring with me. Finally, trying to decide between a good gun (which I can't shoot anyway) and a really good camera (which I can), I take neither, opting instead for a good rock hammer. Not that I think I will need it, but I feel uncomfortable without one, a bad habit that paleontologists quite often pick up.

I decide to wander along the shoreline, and soon any doubts about my temporal coordinates are dispelled. The brushy vegetation alternates with grassier glades and forest in this region, and soon I begin to see game in abundance: eland, a few springbok, wild pig, and numerous black rhino, all belonging to species common in Southern Africa in the Age of Industry, if not always found in the present day Cape region. But *strange* creatures are here as well: a giant horse marked very differently from the familiar zebra and belonging to a species I presume to be the long-extinct giant cape horse; near it a half-horse, half-zebra, surely a Quagga, a species that will survive until the 19th century. I see other novelties as well, such as long-horned buffalo and giant hartebeest, both once common in this region but extinct for perhaps 40,000 years before our time.

Tired, hot, and now quite hungry, I climb a steep cliff of what I assume to be Ordovician-aged sandstone and arrive on a grassy terrace overlooking the beach, now 50 feet below. I am tempted to take a look at the rocks in search of a few good brachiopod fossils, but remembering the mission at hand, I begin to survey the scene about me. I have arrived on a plain rich with game. Several elephant, seemingly identical to those in Africa today, are wreaking havoc among the lowest branches of trees in the distance, and closer to my vantage point I see a group of large, fiercely tusked wild pig rooting in the dark earth, whether warthog or bushpig I cannot tell. They look mean in any case. A herd of deer-like eland is grazing peacefully nearby; a new pang of fierce hunger causes me to look at them not as beautiful African wildlife but as meals. And it is soon clear that I am not the only human on this ridge thinking about eland for dinner.

Coming down the terrace slope I see a band of humans, all carrying spears. I wonder about the prey that has attracted this stealthy group. Could they be the first elephant hunters, or are they after less lethal game? Excited, I grab my binoculars and take a first look at a distant ancestor, wondering if I am seeing a late group of *Homo erectus* or the first true *Homo sapiens*. One glimpse shows these hominids to be the latter, for I am looking in a mirror. I am 115,000 years in the past, on a grassy plain in southernmost Africa overlooking the Indian Ocean. Surely I should see signs of evolutionary change that occurred during that great gulf of time; surely they should look more *cave man* than I. Yet the people I see—the faces and bodies represented in the small band of hunters approaching—seem absolutely, perfectly human in all physiognomic ways. Niles Eldredge and Stephen Jay Gould are right: Evolution for this species—our species—is already largely finished, at least as far as major anatomical change is concerned. Once we speciate, we have completed most of our major morphological change, as predicted by the theory of punctuated equilibrium. The species I see, the first *Homo sapiens sapiens*, is perhaps only 5000 years old at this time, but already its members have assumed the form that they—that *we*—will always have, save for slight tweaks in ornament such as skin, hair, and eye color. They in no way resemble our iconic image of cave men, even though they probably do inhabit one of the large seashore caves in the region. There is no bulging brow or bandy-legged gait; their jaws might be slightly more robust than mine, but that may be my imagination. They wear no clothes or shoes, but each carries a skin bag tied around the waist. Apart from the fact that they are naked, if I ran into any of them on a street corner, I would certainly walk by without pause, and on any French beach in summer there would not even be that excuse to notice. They wear no personal adornment of any kind, but they are all armed. They are black-skinned and finely muscled if somewhat scrawny; in fact, a closer look suggests that these people are not well fed. They carry long spears tipped with stone spearheads, and one has a hand ax. Yet the spear shafts are very crudely shaped, and their stone flake tips seem even more primitive than the arrowheads I made as a child.

From their stealthy, nervous approach, and the wide berth they give both elephant and pig, it is clear that these hunters want no part of either; they are making a direct attack on the eland. With a fierce yell the half-dozen hunters break into a run as they near the herd, charging hard now at the eland, which scatter in all directions. One hunter succeeds in goring a very young eland, but the juvenile manages to flee, trailing a long spear out of its wounded side. Several of the other hunters throw their spears, but all miss, and I begin to see why these hunters are so scrawny: They seem incredibly inept with their spears. By chance one of the hunters stampedes a gray-muzzled eland toward the cliff edge and, with a thurst of his spear, causes the terrified animal to dash over the precipice and fall in a long, graceful arc onto the rocks below, which it hits with a dull concussion. By now the other hunters, winded, have arrived at the cliff's edge, where they all gaze down proudly at the kill below. They begin to clamber down the rocks toward their dinner, their relief showing amid smiles and good-natured hooting, or perhaps talking. There will be food in the Klasies River Cave tonight.

. . . .

I have reset the time machine for 35,000 years ago in the northern coastal region of Australia. And I have brought along a hot air balloon, for I want to see the entire length of this ancient continent, from its southeast shore near what will be Sydney, to its far northern shore along the merging of the Pacific and Indian Oceans.

This time period, unlike that of my South Africa visit, corresponds to a glacial maximum, when the glaciers, which had expanded and contracted across North America over the last million years, were spread to their greatest extent. The great northern hemispheric continents are deeply gripped by glacial ice. So much ice has formed that the sea has dropped as much as 400 feet below its modern-day level. This dramatic drop has linked major land groups long isolated by the sea. In the northern hemisphere, it would be possible to walk from Asia to Alaska if only you could get over the vast sheet of ice covering the land. England and Scandinavia are completely connected. In the south, Australia, Tasmania, and New Guinea are linked in one large continental block we will call Sahul, or Greater Australia. To the north of this great continental mass, the islands of Java and Sumatra are united with the peninsula of southeast Asia to form the Sunda shelf region, thus extending the land area of Asia enormously. Between the Sunda shelf and Greater Australia, a smaller island area called Wallacea acts as a stepping stone between the two. During this interval, the Australian block can be reached from the Asian block by a sea voyage of only 50 miles. It is clear that early humans soon exploited this narrow channel, in the first truly epic sea voyage of our species's history. Arriving in Australia, they found a land filled with beasts strange beyond belief.

I set up the balloon near the coastline of what will someday be Sydney and catch southeasterly breezes for my voyage to the northwest. I hope to land somewhere along the northern coast of Australia.

In this warm afternoon, it is hard to believe that so much of the world is gripped in ice and cold. Yet the clearly extended coastline bears little resemblance to the Australia I will know. There is no Sydney Harbor, no Manley or Bondi Beach with legions of surfers patrolling the shore. As the balloon rises, however, I begin to see a menagerie of Australian animals with appearances skating between the bizarre and the ridiculous.

The low-lying balloon and my strong binoculars bring into focus the Ice Age Australian fauna. Although the eucalyptus trees fringing the coastline are familiar friends here in their ancestral home, many of the other creatures are strangers. Largest and most bizarre are the Diprotodons—lumbering, four-legged creatures as large as a rhinoceros. They are present in staggering numbers and look like dim-witted nightmares from an H.P. Lovecraft book. Small eyes peer myopically from massive heads as I skim overhead. Yet perhaps the most bizarre feature of these strange animals is their relatively tiny feet: The giant Diprotodons seem to mince rather than walk. They seem incapable of rapid movement—not a serious disadvantage as a result of their size and the virtual absence of marsupial predators large enough to do them much harm. Moving among the huge Diprotodons are other large, lumbering creatures, all marsupials. Giant kangaroo hop about: *Macropus ferragus*, the largest kangaroo of all time. Other, slightly smaller kangaroo mix among them, but these, belonging to the genus *Procoptodon*, are far more bulky than the *Macropus* and must weigh as much as 700 pounds. Their heads look every bit as though they were transplanted from a rabbit and then grown huge. They hop, all right, but in a peculiar, awkward fashion. Speed is clearly not the top priority of any of these great beasts, all now long extinct.

My balloon is flying inward across the continent, and soon the verdant coastline gives way to drier grassland and eventually to desert. Occasional ponds, rivers, and small lakes appear, and around these oases game is again abundant. I see many more kangaroo and wallaby, as well as giant wombat, which are ground sloth-like creatures, and a variety of fierce-looking carnivores that look like either cats or dogs or sometimes resemble a combination of the two.

Drawing of Diprotodon from ancient Australia. (From Martin and Klein, eds., 1989.)

By now I am well over the interior of the continent, and most game has disappeared. A blazing sun and the rocks skimming by below are my only company. I keep trying to tell myself that it is the heart of the Ice Age, that far to the north great woolly mammoths roam in the lee of mile-high walls of ice, that the land is gripped by winter, and that my species, our species, is only now colonizing the northern reaches of Asia. The great trek toward North America, across the Beringia land bridge, is still lies 20,000 years in the future. But the heat and sun suggest otherwise; the Ice Age, from this perspective, is nothing like the myth embraced by my distant world.

I am moving swiftly toward the northern coast now, heading across the surprising flatness of Australia, passing over ancient dry ground largely undisturbed for millions of years, catching sight of the occasional kangaroo and sometimes spotting creatures much stranger. Here and there I see giant lizards, bigger than anything on earth in my day, varanid monitors allied to the Komodo dragon of Indonesia but much bigger, easily large enough to eat a kangaroo—or a human, if any should be about. Yet although I have now flown over great expanses of this huge continent, there is still no sign of humanity among the throngs of Australian wildlife.

The balloon floats northward, and the heat is almost intolerable. But I know that my continent-crossing journey must be nearing its end, for the trees begin to reappear. Giant marsupial game becomes common again, and finally, far in the distance, I see the low blue expanse of the ocean. I am alert now, and the balloon is rushing forward. As I approach a beach a new vision appears: In the ocean is a poor sort of raft, logs lashed together but clearly battered, and on that raft sit dark-skinned men and women, weakly paddling. The raft makes shore, and the humans tumble off. Some lie starving on the sand, while others, carrying stone-tipped spears, look with more than passing interest at a large herd of slow-moving Diprotodons watching them from the shoreline.

. . . .

It is the last voyage, 11,000 years into the past. I have arrived on a rolling plain, and my first impression is of the strong wind, howling, carrying gritty sand. The day is hot, and although I have traveled to somewhere in the southwest of North America, this landscape might as well be East Africa. Thin clouds scud overhead as I crest a low, grassy hill to face a vast expanse of prairie. No buildings, no roads, no trees; a few dried flowers and shrubs are interspersed, but mostly I see this grassland under blazing sun, and a great dryness. Yet there must be more to see here than yellow grass; there must be animals. So I begin to wander.

In the heat I soon rue the lack of water. Far to the north of me there are still some isolated glaciers, but on this scorching day, the knowledge that the planet is emerging from the Ice Age is little solace. Believing that so much of the continent has recently been locked in glacial ice is difficult on this day of remorseless sun amid the parched, yellow grassland.

After an hour of wading through the dusty fields, having seen only a distant herd of bison, I begin grumbling about all those learned scientists who have conjured up images of the game-rich prairies of North America's Ice Age. Lions and tigers and bears, yeah right, and of course as soon as I say this, I stumble on more than seem possible.

I have come over another anonymous hill, but below me is not just another grassy slope. A new panorama emerges: a large, muddy flat with small pools of water in its center. Within this great wallow stands a herd of mammoths, while giant condors wheel overhead, watching. The mammoths are huge beasts, up to 13 feet at the shoulder, with the knobby head and sloping back that are characteristic of their kind. Yet they have very short fur, and the gently inward curling of massive tusks shows that this species is *not* the familiar woolly mammoth of more northerly regions but the Columbian mammoth, a species adapted for the broad, warmer grasslands of North America's Great Plains. Perhaps 20 to 30 of the beasts, large and small, are congregated in the wallow, and it is soon clear that they are not in good spirits.

I move closer and see that the larger animals have excavated great burrows in the soft mud and are drinking from the brackish, black water that oozes into these trenches. Others of the group use their massive, inward-curving tusks or shuffle their feet to dig new pits elsewhere on the broad flat. This region is surely a large pond during the rainy season, but now it is mostly mud and little water. Smaller members of the herd try to enter these crude excavations as well, but the larger mammoths shoulder them away. This callous behavior seems incongruous; I have always thought of elephants as excellent parents. Nevertheless, the adults have walled out the youngest, and the cries of these thin, bony baby mammoths, clearly dying of thirst, are pitiful. It is also clear that the newly dead were not the first to die here. Numerous large, white bones stick out of the soft black earth around the seeps, and from time to time one of the larger mammoths walks upon a newly uprooted bone, crushing it.

There are also other objects here in the mud, and I walk closer to investigate a familiar shape. A broken spear tip lies amid some crushed bones, clear evidence that the first Americans, whom we call the Clovis People, know there is meat to be found here. For the first time I get the feeling that as I watch the mammoths struggle, I myself am being watched by unseen human eyes nearby.

I move in closer still to the mammoths, and none of the great creatures pays me much heed. One of the younger animals has collapsed. I walk up to it, touching the brown fur, looking at its terrified eyes as it gasps in the hot sun, and it is clear that with a quick spear thrust through the heart and a fast evisceration, I too could be a mammoth hunter.

. . . .

Three stories about the past, not even good science fiction. But the time machine analogy is powerful because it lets us conduct "thought experiments." I am neither the first nor the last time voyager. But for me, these Ice Age expeditions are a novelty, my usual haunts being much farther in the past, at places where other, older, and even more catastrophic great extinctions can yield clues to the fate of the Ice Age mammals. If the mammoths and mastodons are the dinosaurs of the Ice Age, then perhaps it behooves us to visit the *real* dinosaurs and look in more detail not only at what killed them off but also at how this sudden dying is preserved and recorded in the fossil record.

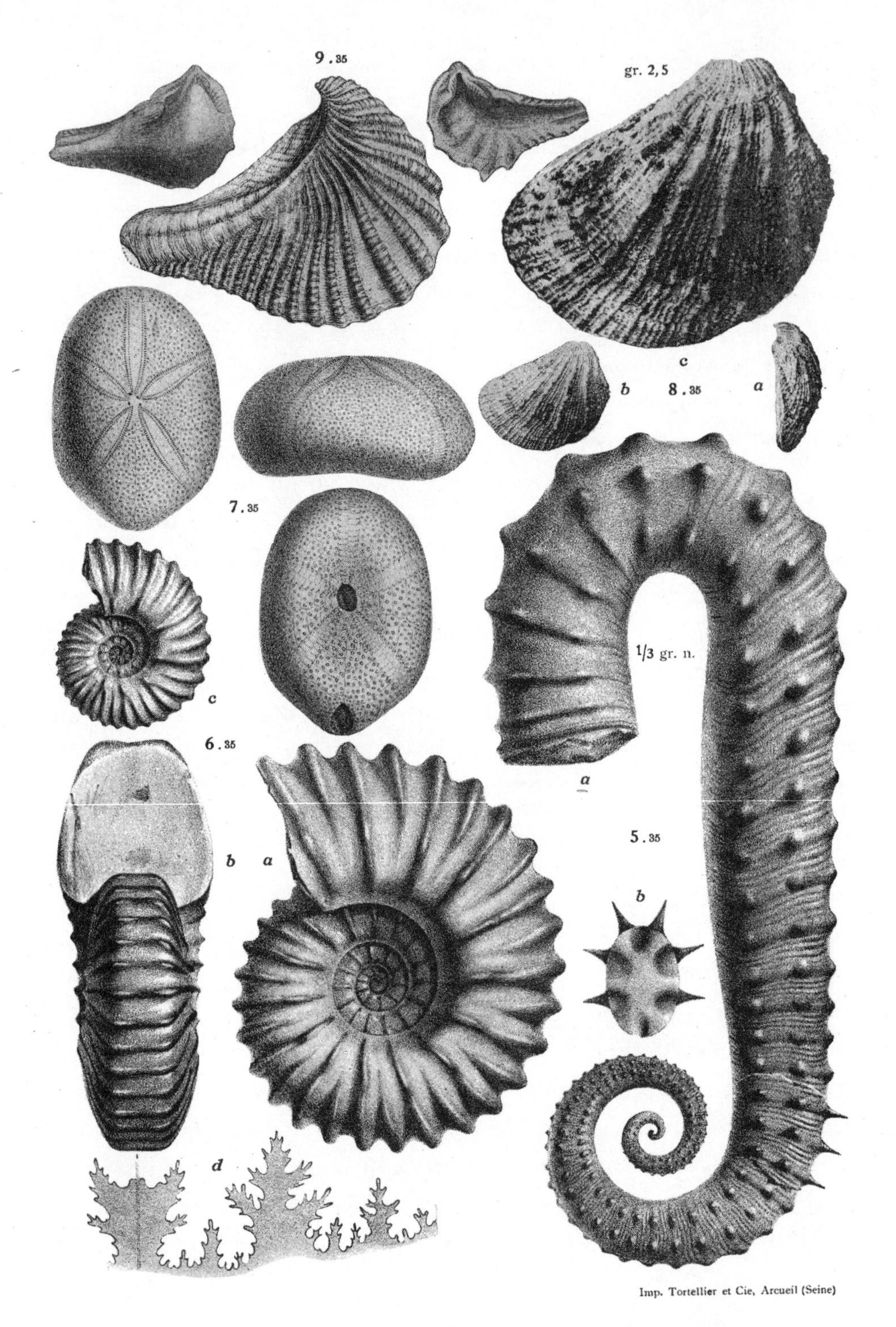

PLATE OF FOSSIL SHELLS FROM FRANCE. THE SHELLS' SUDDEN DISAPPEARANCE HELPED CONVINCE CUVIER ON THE REALITY OF CATASTROPHIC EXTINCTION.

2

Heart of Darkness

WHAT, OR PERHAPS *WHO*, was the cause of the Ice Age mammal extinction? This question has puzzled scientists since the turn of the 18th century, for the dying-off of large mammals at the end of the Ice Age was far more than a minor convulsion of Earth's biosphere. It was a mass extinction.

A simple species extinction is spooky enough. A species consists of organisms capable of interbreeding; for an entire species to go extinct, therefore, every last member must die. Extinction of a single species usually has little effect on the planet's biota. (*Our* extinction would certainly be an exception to that rule.) Occasionally during the last 500 million years, however, a significant proportion of all species living on earth went extinct within a short span of time. These events had profound effects on the evolution of other species. For example, had the dinosaurs not gone extinct, there is little chance that there would ever have been an Age of Mammals, or an Age of Humans, for that matter.

About 15 of these mass extinctions have occurred in the past half-billion years. Five are considered "major" in the sense that more than half the species then on Earth disappeared in several million years or less. Two of these, the event that killed the dinosaurs and an earlier event that ended a time interval known as the Paleozoic Era, are the most relevant to understanding the Ice Age calamity. Like the Ice Age mass extinction, both of these earlier events gravely affected terrestrial animals. They also created the largest time units recognized by geologists, for these two most lethal of mass extinctions of the deep past so changed the nature of life on Earth that they are used as the major time markers of the geological record. They are at the heart of the geological time scale.

The geological time scale is composed of nested, hierarchical units. Understanding these units, so familiar to all 10-year-old dinosaur aficionados (but so inaccessible to the rest of us) is absolutely necessary in studying the past. The longest are called eras. Eras are separated into shorter units known as periods, which are in turn divided into epochs. Paradoxically, each unit of the geological time scale was originally defined in terms of fossil content only, rather than measured in numbers of years. Only recently have we been able to attach numerical ages to the boundaries separating these geological time scale units by dating rocks and fossils using isotopic techniques such as argon and radiocarbon dating.

The time scale was formulated over many decades of the early 19th century, long before scientists had the ability to date rocks in terms of their absolute age. The motive in erecting the time scale was to sort the immense, seemingly chaotic piles of sedimentary rock on the earth's surface into chronological order. Because only fossil content made this possible, and because significant changes in fossil content were most often caused by mass extinction, the mass extinctions provided the punctuation marks in the great stratal book of geological and biological history on our planet.

Even the most refined unit differentiation was based on the recognition that different assemblages of sedimentary rock often contain differing fossil content and thus record different time intervals. Because sedimentary rocks, the bearers of fossils, are originally laid down one stratum at a time, one literally moves up through time by moving up a sedimentary rock column. The arcane names still used in the geological time scale are based on the regions where these units were first recognized. The periods known as the Cambrian, Ordovician, Silurian, Devonian, Mississippian, Pennsylvanian and Permian constitute the Paleozoic Era, the "time of ancient life," which dates from 530 to 250 million years ago. The Triassic, Jurassic, and Cretaceous make up the Mesozoic Era, the "time of middle life," which extended from 250 to 65 million years ago. The Tertiary, Quaternary, and Recent are the periods of the Cenozoic Era, the "time of new life," which extends from the end of the Mesozoic to the present day.

Because successive units could be recognized only by their differing fossil content, the boundaries between units were drawn at the points where the rock strata show evidence of some convulsive faunal and floral turnover. The greater the change, the higher in the hierarchy the boundary. (That is, changes that separate epoch are much less dramatic than those that separate periods, and the differences between eras are the greatest of all.) One of the great pioneers of this process, John Phillips, recognized in 1860 that mass extinctions are the single most important factor leading to a change of fossil content within strata and that they were therefore of the greatest importance in establishing the time scale. Phillips proposed that the two most catastrophic of the mass extinction events, one ending the Permian Period (now known to have occurred about 250 million years ago) and the other ending the Cretaceous Period (65 million years ago), separate the last 500 million years of Earth's history into three great intervals, each with its own distinct suite of marine and terrestrial fossils. It was Phillips who proposed the names of the Paleozoic Era, whose end is marked by the Permian–Triassic mass extinction, and of the Mesozoic and Cenozoic Eras, which are separated by the Cretaceous–Tertiary mass extinction. We still use these names today.

Because of their importance in subdividing the time scale, it is no surprise that these two greatest of mass extinctions are the best studied. The

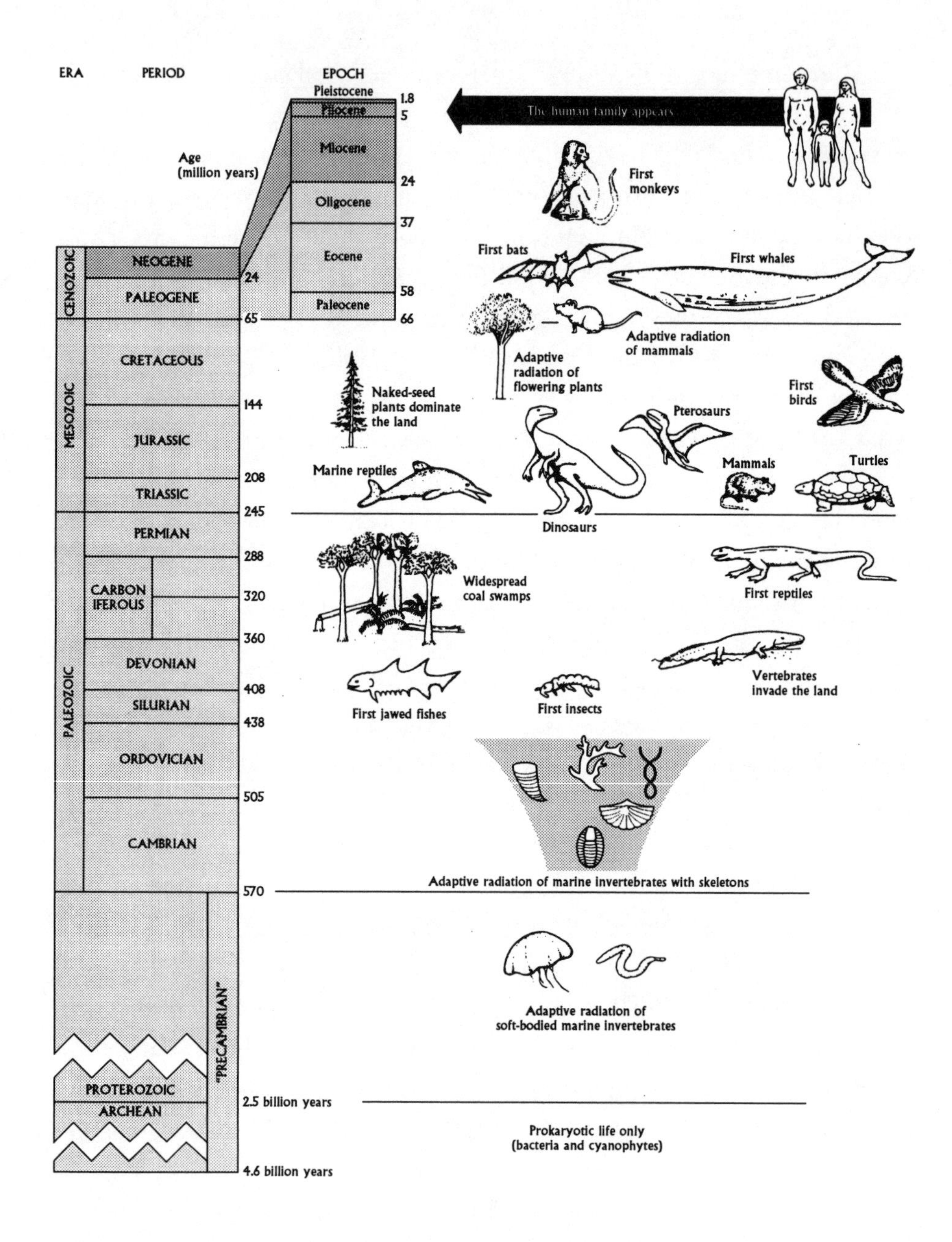

Geological time scale. (Modified from Steve Stanley, Earth and Life Through Time; by permission of WH Freeman.)

older of the two, the Permo-Triassic event of 250 million years ago, is acknowledged to have been the more consequential, involving the death of as many as 90% of all species then living. Yet because of its greater antiquity, the Permo-Triassic event is less well understood than the Cretaceous–Tertiary, or "K/T," event of 65 million years ago.

The K/T event resulted in the extinction of about 60% to 70% of all species then on Earth, including, most notoriously, the dinosaurs. This great cataclysm was originally thought to have been caused by exploding volcanoes or changing climate and thus to have taken millions of years to complete. Since 1980, however, we have developed an entirely different view of both the cause of this extinction and the time it took. Most geologists now believe the K/T event was brought about by environmental perturbations following the sudden impact of a comet 10 miles in diameter hitting the earth. The consequences of that blow—acid rain, rapid temperature change, global forest fires, months of darkness and the suppression of photosynthesis—created a short period of mass death lasting from a few tens of years to perhaps a few thousand. When the dust literally settled, the dinosaurs and many other of the earth's land and sea species were extinct. Many more followed in the next millennia as large-scale atmospheric perturbations poisoned the earth.

The discovery that the dinosaur-killing mass extinction was caused by a comet or asteroid striking Earth was made in 1980 by a Berkeley group that included the father-and-son team of Luis and Walter Alvarez. The Alvarez impact hypothesis, as it is now called, was originally controversial, but mineralogical, chemical, and paleontological data gathered during the 1980s persuaded most scientists familiar with the problem that a large comet or Earth-crossing asteroid did indeed hit Earth approximately 65 million years ago. The discovery in the Yucatan region of Mexico of a large, 65-million-year-old, multi-ring impact crater (now called the Chicxulub Crater), about 200 miles in diameter, has largely swept away remaining opposition to the hypothesis. We are now virtually certain that the end of the Mesozoic Era was brought about by the impact of one of the largest extraterrestrial bodies to crash into Earth since formation of the lithosphere.

Some thought at first that all mass extinctions could be explained by celestial impact. In the 15 years since the Alvarezes' discovery, many of the other mass extinctions have been investigated to see whether they, too, could have been brought about by unwelcome visitors from the edge of the solar system. To many people's dismay, however (and to the delight of many more), virtually none of the other mass extinctions shows evidence of having been caused by celestial impact. A variety of other causes have subsequently been found, including climate change, change in Earth's atmospheric gas concentrations, and excess volcanism. For the most recent mass extinction, the Ice Age calamity that wiped out so many large mammal species, the presence of humans may have caused a large part of the damage.

A great irony of the Ice Age mass extinction is that even though it is the most recent of the 15 mass extinctions that have occurred over the last 500 million years, it is far from the best known or most thoroughly studied. Yet generalities do emerge, and lessons from one event can often be used to study the others. The Permo-Triassic and K/T events have been especially pertinent to our understanding of the Ice Age event, because all three saw the rapid extinction of the majority of large land animals. The methods of studying how large animals go extinct unify these three catastrophes. It is from the study of the Permo-Triassic and K/T extinctions that we can best gain new nsights into the Ice Age calamity.

Uncovering the killer of the great Ice Age faunas requires numerous tools: theory, which provides a list of suspects and perhaps motives; field work, which produces data; and the data themselves, which will ultimately solve the mystery by enabling us rule out various hypotheses. The most critical data, and by far the most elusive, are those that yield accurate information about time. We need to know whether the Ice Age extinction took hundreds, thousands, or many thousands of years to unfold; we need to know whether it had one major cause or many. Both of these factors are best approached by trying to pinpoint the time involved in the event.

A starting point is also required. In most whodunits, the adventure starts at the scene of the crime. In this case, however, we are dealing with mass murder, global in distribution; our inquiry perhaps best starts in the place where the crime was first discovered. For that, we need to travel in our time machine to a formal garden in Paris, where an ancient museum packed with yellowed, moldering bones celebrates the life and work of George Cuvier, the father of comparative anatomy and the man who first recognized that species extinction is a reality. Cuvier, contemporary and confidant of Napoleon, was the first vertebrate anatomist and perhaps the greatest of all time. He was also the first specialist on mass extinctions, and he realized that to understand any single mass extinction, one needs to study them all.

In September 1995, I began such a voyage to Paris, but as it turned out, my route was by no means direct. My detours were caused by scientific meetings, one dealing with the Permo-Triassic mass extinction and a second with the K/T event. At the start of this voyage, I had hoped to close out a decade of study of these two older extinctions before launching research into a newer mass extinction, naively believing that the Ice Age mass extinctions could be understood as being qualitatively and quantitatively different from the older events. This was a mistake; like Cuvier, I now believe that truths learned about global biotic catastrophes of the deep past are necessary tools for understanding the more recent catastrophe that decimated the larger mammals of the Ice Age. Such understanding may even be indispensable if we hope to preserve biodiversity on this planet.

. . . .

I hate the frantic period just before departure on an intercontinental flight. Paying bills, working on manuscripts up to the last minute, preparing talks, buying cat food, packing bags as lightly as possible (bags that inevitably turn out to be far too heavy); cramming in clothes for a month—and for a variety of occasions. I'll need clean but casual clothes for my first stop, a conference in Washington, DC, to learn and talk about the cause and biological effects of the Permo-Triassic mass extinction in informal circumstances amid the great displays and collections of the Smithsonian's Natural History Museum.

Formal clothes are needed as well: After Washington I'll face a week-long conference in Brussels, where 200 scientists will discuss the opposite end of the paleontological spectrum from the Washington agenda. At Brussels, the topic will be time, not biology. This meeting will be concerned with establishing the rigorous chronological time lines for the Cretaceous Period, the last interval of time of the Mesozoic Era and the time unit that ended with the dinosaur-killing mass extinction. The very formal Europeans who convened and numerically dominated the meeting would wear suits and ties, smoke pipes, have little professorial beards. They would have little sense of humor about their job or the ways of their more casual American cousins. And they certainly would have little sympathy for biological theory; to them, paleontology is an applied science.

The suitcase then received far more comfortable denim and cotton attire and heavier gear as well—the cold steel of hammer and chisels, the burnished leather of field belts and cases, precision compasses and measuring tapes, cameras, satellite global positioning receivers, blank specimen bags, and magnifying loupes; all the totems of the geologists' guild were interlayered with the stratified clothes. Field collecting gear would be necessary for my chosen road from Belgium, for I intended to travel south out of the Low Countries, through Luxembourg, France, Switzerland and the Alps, to Italy; then over the Apennines nearly to the Adriatic Sea, where stony gorges and high stratal walls exposed in the ancient mountains near Gubbio, Italy, are the site of what is arguably the most famous geological outcrop in the world.

It was at Gubbio that a thin platinum and iridium layer imbedded between Mesozoic and Cenozoic strata was first discovered by the Alvarez team—a discovery that eventually convinced the scientific world that dinosaurs, mosasaurs, ammonites, pterosaurs, and the other denizens of the Mesozoic Era, or the Age of Dinosaurs, had been destroyed by the impact of a great comet. My work at Gubbio will be directly related to this mass extinction: I want to look at events that occurred in the several million years immediately prior to the meteorite impact, when this part of Italy was far to the south of its present position, when it was a deep, quiet seabed instead of a rocky landscape. I intend specifically to search for evidence that the great mass extinction that ended the Cretaceous was more complicated than even the Alvarezes have suggested and was not caused by one single event.

Last, with the bag crammed to bursting, I squeezed in a simple blank notebook to hold scribbled thoughts for the last stage of my journey, in

Paris, to finally visit the shrine to Cuvier, the man who proved the reality of extinction.

As I gathered and packed my things, I mulled over the long trip. Before the start of this odyssey, there came a moment of inertia. It dawned on me that it was not the trip I found so daunting but the prospect of writing this book and, in the process, perhaps coming face to face with yet another ancient murderer. In strata of the not-so-distant past are buried the suspects, murder weapons, death beds, and graves of some of Earth's most magnificant creatures, the great Ice Age large-animal fauna. With these clues surely lies the identity of the murderer. There are only two prime suspects: climate change and our human ancestors. Neither is a very palatable alternative.

If climate change, in ending the last glacial episode, caused the death of so many great creatures, then our species must take a very hard look at the possibility that human-produced climate changes taking place *now*, and into the next century, will set off a second wave of extinction. Even though all but the politically motivated agree that human-produced emissions are rapidly producing a buildup of carbon dioxide and other greenhouse gases, and are probably raising the mean global temperature in the process, no practical steps have been taken to limit such emissions. If greenhouse gases are accumulating now, when the world population is "only" about 5.5 billion people, what will happen in the next century, when our numbers reach 11 billion, many of whom who will depend on coal-fired industries (the leading cause of CO_2 buildup) for their housing, clothing, and livelihoods? It is becoming increasingly clear that human-produced emissions of greenhouse gases have already altered the world's climate in a way directly analogous to what occurred naturally between 15,000 and 10,000 years ago. Then, a giant global catastrophe occurred—perhaps coincidentally, perhaps not. Are we facing the same prospect?

The second prime suspect is humanity. One of the major hypotheses advanced to explain the death of the great Ice Age faunas is that humans, hunting for food, caused extinctions among larger animals whenever we entered a new, previously isolated land mass. The majority of anthropologists believe that humans arrived in Australia about 40,000 years ago, in North and South America about 12,000 years ago, and in Madagascar and New Zealand about 2000 years ago. In every case, extinction of the larger animals in these habitats soon followed. If bands of human hunters, armed only with spears and stone-tipped implements, can cause the extinction of such a large number of great creatures, what will be the fate of our own world's faunas as human populations swell and human parents need to feed more and more hungry babies? The scientist within me relishes a new mystery, yet I cannot help but wonder if some mysteries are better left unsolved.

. . . .

I arrived in Washington sweltering in the heat of a long, rainless summer. By early September 1995, the east had endured nearly 2 months of drought. New York City was considering water rationing, and the weather, or rather the strangeness of it, overshadowed even Newt Gingrich, Colin Powell, and the imminent financial default of the nation as a topic of street-corner conversation. It turned out that the Washington meeting I attended was also concerned with the weather, but in this case the Paleozoic weather of 250 million years ago.

It was an odd sort of meeting as scientific concourses go, being small, with attendance by invitation only. There were several Europeans, a Chinese, and a Russian scientist. But most of us were Americans: Douglas Erwin, David Jablonski, Steven Stanley, and Greg Retallick, among others, all members—or future members—of the National Academy of Science: the finest American paleontologists. All were specialists on mass extinctions or on the Permo-Triassic boundary and its great mass extinction of 250 million years ago.

The gathering in Washington was an attempt to propose new hypotheses to explain this single greatest mass extinction. It was certainly not caused by a comet or an asteroid striking Earth. There is no evidence of such an event, such as the traces of iridium particles, glassy tektite spherules, shocked quartz, or carbonized layers that are now known to be the deadly calling cards of a calamitous celestial impact. The Permo-Triassic extinctions were also assumed to be too chronologically drawn out to have been caused by an impact. This catastrophe is thought to have lasted hundreds of thousands and perhaps even millions of years, instead of the few years of agony—and mass death—that follow an impact event. Scientists were forced to conceive of other explanations, cleverer means of killing the animals and plants, swimmers and crawlers and scaly creatures, and the myriad other life forms that succumbed to . . . what?

And so we convened, coroners really, assembled to dream up new scenarios for the destruction of an ancient biological world and speculate about sudden temperature changes, a sinking sea level, the buildup of poisonous gas, and heavy-metal poisoning. Geologists, who perhaps are more comfortable constructing explanations based on hard fact and observation, found themselves forced to emulate a scientific style much more familiar in theoretical physics by using the *gedanken*, or "thought experiment," made popular by Einstein and others. It was just as well that our thought experiments could not be conducted in reality, for not even the most environmentally oblivious political party would have authorized and funded our dumping tons of sulfur on eastern North America (as might have happened at the end of the Cretaceous Period) to see its effects on the biota, or unloosing all of the world's hydrogen bombs simultaneously to test Carl Sagan's hypothesis about impact winter in the aftermath of a nuclear attack. But one of our thought experiments, perhaps quite as lethal as the others, is in a sense already being implemented. Huge volumes of carbon dioxide are currently being pumped into the atmosphere, and this seems like an efficient cause of a mass extinction. Indeed, a similar event may already have produced the single greatest mass extinction in Earth's long history.

Carbon dioxide is a wonderful thing if you are a plant, but animals don't live long when carbon dioxide levels in the air or water that surrounds them exceed 6% to 7% by volume. Thus, if you were a paleontologist looking for a clever way to kill off a biological world, you might invoke sudden changes in the global gas inventory and devise a plausible method of supercharging the atmosphere with carbon dioxide. And sure enough, at the Washington meeting we first heard about a new theory proposed by paleontologists Andy Knoll and Dick Bambach and geologists John Grotzinger and Don Canfield.

Imagine a world with but one continent and one ocean, both necessarily gigantic. Make your ocean rich in calcium carbonate and dissolved carbon dioxide in its deeper portions, and then suddenly mix the entire brew so that these deeper waters are suddenly thrust to the surface. The net result is the sudden release of immense volumes of carbon dioxide into the atmosphere. The excess carbon dioxide kills both directly, by carbon dioxide poisoning, and indirectly, by creating the so-called greenhouse gases that tend to trap solar heat in the lower atmosphere and thus raise global temperatures.

We have recently seen a dramatic demonstration of the lethal effect that sudden liberation of carbon dioxide can have. In 1986 a deep crater lake, Lake Nyos in Cameroon, Africa, suddenly released a great volume of carbon dioxide from its deep bottom. This sudden gas discharge killed more than a thousand people and untold numbers of cattle along the edge of the large lake; all died of asphyxiation from carbon dioxide poisoning. The release of gas at the end of the Permian Period is not thought to have killed in a manner of minutes, as it did in Africa. But kill it did, in the sea as well as on land.

As if carbon dioxide released from the ocean were not already creating enough misery in the Permian world, a paroxysm of volcanic activity added even more of the gas to the atmosphere. Near the end of the Per-

The Permian extinction episode. (From L. Figuier, La Terre Avant Déluge, 1864.)

mian Period 250 million years ago, there was a sudden onset of flood basalts in what is now Siberia. Flood basalts are watery magma erupted in enormous volumes onto the earth's surface. Concomitant with such eruptions today is the release of huge quantities of CO_2 into Earth's atmosphere; we assume that CO_2 was released on a far greater scale in the late-Permian eruptions. Volcanism and the release of dissolved CO_2 from the ocean at the end of the Permian Period may have combined to cause the greatest mass extinction yet to have occurred on Earth. Thus, a quarter of a billion years ago, animal evolution was almost reset back to the simplicity of pond creatures: hydra, protozoa, and *Daphnia*.

After this great Permo-Triassic mass extinction, the few survivors found themselves in a world of very few competitors and very few predators. As the deadly CO_2 diminished, the surviving animals and plants rapidly multiplied, and new species evolved and began to repopulate the world. But the mix of flora and fauna was decidedly different from what had been dominant before. (This seems to be a general condition of mass extinction: In the aftermath, as after a forest fire, the same mix of animals and plants never returns.

There may have been one key to surviving on land: warm-bloodedness. Although this theory is still unproved, two groups of reptiles living at the time may have been warm-blooded: the ancestors of mammals and the ancestors of dinosaurs. Many other reptiles lived on Earth at the time, but nearly all of them were wiped out. In the competition for dominance of the terrestrial ecosystems following the Permo-Triassic extinction, the first dinosaurs won out over the first mammals. And many paleontologists think that mammals would never have achieved dominance were it not for another chance event that occurred when, 65 million years ago, an incoming comet changed the rules again.

The Washington meeting was a watershed in at least one way: For the first time, sentiment shifted away from explanations that favored a gradual, drawn-out extinction to something far more catastrophic. The old theories about this particular extinction—long assumed to have been the least rapid of all mass extinction—gave way to more abrupt scenarios much like our explanations for the other extinction events. It appeared that many of my colleagues were becoming adherents to Catastrophism, which holds that major episodes in Earth and life history came about through sudden catastrophe, not slow progress. The results of the meeting were reported in the popular media—National Public Radio, *Time* magazine, *The New York Times*, and *National Geographic*. But the real progress came about through communication among the various paleontologists assembled.

Our meeting ended on an unexpected, somber note. We learned that a team of government meteorologists had just announced the results of a new computer model simulating the amount of global warming produced during this century. They found that it was rising rapidly. They also warned of what this could mean to humanity and the rest of Earth's biota: With increased global warming, we can expect partial melting of the icecaps, causing a rise in sea level; an increase in the size of deserts; and a vast increase in the frequency, intensity, and duration of storms such as hurricanes and typhoons (a heated atmosphere produces more storms than a cooler one). Perhaps mass extinction as well? Soon after this announcement, a legislator from California suggested that the concept of global warming was nothing more than "liberal claptrap."

. . . .

In Washington, the agenda had been dominated by paleontologists interested in understanding the biology of an ancient event; the exchange of information was mediated by the easy informality of Americans all well known to one another. In Brussels, 200 paleontologists of a very different stripe gathered in the ancient Natural History Museum for a different kind of work. The Brussels meeting was dominated by Europeans more formal in approach and mission. Time, not biology, was the subject of the meeting—not time as it may be enlisted in an effort to answer comprehensive questions about the history of life, but time as a concept devoid of larger meaning. It was a meeting about *biostratigraphy*, a study that dates back to early in the 19th century, when the great German and English professors struggled to construct the geological time scale, unconcerned with any possible application. The nature of paleontological science has changed since then. The great excitement in paleontology now stems from its application to larger questions about the history of biodiversity. A passage from an old, classic movie came to mind as I watched the European-dominated assembly debate the least nuance of geological time with no concern for how the concept might be utilized. It reminded me of Alec Guinness building the largest, longest-lasting bridge possible over the River Kwai without giving any thought to what would cross the bridge or the uses to which it would be put.

This conference had gathered for a periodic re-examination of the time lines used to subdivide the Cretaceous Period of 135 to 65 million years ago. After 200 years of study, the boundaries between geological time units can be modified or changed only through international consensus. Clearly, a time scale cannot be controlled by local politics. The study of the past requires a time scale that transcends international boundaries as well as temporal ones. Once put into service, the time scale may be adjusted only at risk of changing the entire edifice. Thus, elaborate international commissions whose sole function is to impede progress have been in place for more than a century. But progress in geology does occasionally necessitate changes to the time scale, especially now that new, more modern methods for dating rocks are being employed.

Brussels meeting featured a clash of cultures between the old school of biostratigraphers, who rely on nothing more than the fossil record as retrieved from the rocks, and the newer generation of paleontologists and stratigraphers, who rely as much (or more) on methods unimagined half a century ago: carbon-14 dating; unbelievably precise argon–argon dating, which takes single crystals, vaporizes them with laser beams, and then reads the age of the crystal and hence of the rock it came from; magnetostratigraphy, wherein the periodic reversals of Earth's magnetic field, preserved in strata through orientation of minute particles of ferrous mineral grains, is read from the rocks with gigantic magnetic machines; isotope stratigraphy, in which mass spectroscopes read the ratios of oxygen and carbon isotopes to glean information about the ancient temperatures and productivity cycles of long-forgotten oceans. All of these laboratory-based practices are revolutionizing geological dating, and they are increasingly making the lone geologist of yore, slowly collecting and measuring strata, an anachronism. The changing of any guard, the displacement of great thinkers and bygone methods, is always painful.

Some of the most interesting things I learned in Brussels came from practitioners of these new isotope dating techniques. The revolution in our ability to date rocks (which was impossible until the 20th century and replete with error until the last several decades) has in many cases completely changed our view of the past. I was intrigued by the latest techniques, not so much for what they said about the Cretaceous Period but for their implications for the Ice Age extinction of large animals, which was much on my mind. The greatest debate between various researchers into the mystery of the Pleistocene elephant extinction is related to carbon-14 dates, for even though most carbon-14 dates suggest that the extinctions took place over a short period of time, some researchers have suggested that many carbon-14 dates recovered during the earlier years of the method's use are in error. New methods have called into question many dates produced more than 20 years ago, dates that are absolutely crucial in identifying the killer of the Ice Age fauna. There was no shortage of emotion among the various scientists defending their older findings, which of course had once been on the cutting edge, or among the young turks intent on slaying the old rulers and taking their place.

. . . .

After a week in Brussels I took a long train ride south, glad to be free of the gray city. I looked forward to a sunnier Italy, to the September heat when the grapes are harvested and the tall sunflowers lining the countryside are shorn. Instead, I moved through a landscape devastated by unceasing rain, with swollen rivers, rotting vineyards, and blackened, flooded fields. As great typhoons battering Japan and furious hurricanes smashing the Caribbean were reported by radio, I crossed a southern European landscape inundated by the worst summer rains of several lifetimes. Against this unsettling backdrop, European scientists echoed the earlier American finding that significant, human-produced global warming had been confirmed and that the net effect would be shifting, unpredictable weather patterns and more violent and frequent storms.

My journey to Italy was a pilgrimage to see the Gubbio gorges, sites of the first discovery by the Alvarez team, in 1980, that the Age of Dinosaurs ended suddenly in violence, fire, and biotic catastrophe some 65 million years ago. Accompanied by one of the principals in that discovery, Allesandro Montanari (a former student of Walter Alvarez), I spent several tranquil days looking at thick sequences of limestone marking the millennia of the dinosaurs' empire and the epochs of aftermath following that empire's collapse.

I learned two lessons during my Italian stay, and though neither had much to do with the science I pursued, both told me a lot about how science works. First, a fresh perspective can lead to new ways of seeing things and often to new discoveries). Second, as in scuba diving and parachuting, the risk of fatal errors is greatest both very early in one's career and very late (early because of lack of experience, late because of the complacency and ego that success and familiar routines can breed).

For 15 years I have slowly chipped new information about the K/T mass extinction from seacoast rocks in the Basque country of France and Spain. There, the record of the last years of the Age of Dinosaurs is now quite clear: A sudden extinction of unprecedented rapidity occurred. But the record is more ambiguous if you look at the way marine fossils in those strata undergo extinction by examining not just the rocks that record the last several thousand years of the Cretaceous Period but also strata spanning the period's last 2 *million* years. This wider view makes it clear that the end-Cretaceous extinction was not triggered by one sole cause. Although the greatest single component of the extinction did indeed occur 65 million years ago and was certainly related to the gigantic impact event recorded as the Chicxulub Crater in the Yucatan Peninsula, an earlier extinction took place about 67 million years ago. In the sea, this is shown by the disappearance of reefs and of large flat clams of a type unknown today, clams as much as 3 feet across. These curious creatures, called inoceramids, lived in most ocean bottoms, from the very shallow, sunlit parts of the sea to the deep abyss. They died out long before the meteor hit, so this particular extinction cannot have been related to the more extensive extinction that occurred 2 million years later.

The implication of this observation is clear: The Cretaceous–Tertiary extinctions were brought about by at least two independent causes. This raises some interesting questions. For instance, had the first extinction event not occurred, would the subsequent extinction caused by the comet's impact have been so destructive? And if the comet had not hit Earth, would the earlier extinction have been noticed these millions of years later? I believe the two events, only 2 million years apart, acted in concert to amplify the overall extinction to a greater level than either, alone, would have inflicted.

It is my hunch that all of the great mass extinctions were similarly multicausal, undoubtedly involving different causes in each case, but always the result of a variety of killing processes. On my journey to Italy, I wanted to see whether these giant Cretaceous clams were also found in the Gubbio strata and, if they were present, whether they too disappeared prior to the record of the impact event.

Unfortunately, in the thousands of words written about the Gubbio strata during the last 20 years, not a single paper mentions clams (or any other fossils larger than single-celled protists) in these beds. Indeed, because the thick limestones making the Apennines were deposited on a sea bottom more than a mile deep, it was assumed that no larger fossils would *ever* be found. Very few larger creatures capable of leaving a skeletonized fossil inhabit such environments today, so it was considered doubtful that such large creatures could have lived in the deep oceans back then, either.

We took an Italian beater of a car through the steep gorges, lurching at the breakneck speed that seems to be the norm on Italian highways, into ever more rugged mountains, arriving finally at a high mountain pass through stacked porcelain limestones. My friend Sandro had spent his professional life working in these rocks, and he did remember seeing a few rare clams; to our surprise we found large numbers of them, heretofore unnoticed, at the same stratigraphic interval in which I had seen their remains in the distant cliffs of France and Spain. As the days went by, we visited more sites on more Italian cliffs, and the result was always the same. Over a 10-meter interval the clams were common, and then they disappeared as we crawled upward through the strata, and up through time. At the same level in each section, 2 million years prior to the iridium-rich layer, they vanished, like the rest of their kind elsewhere in the world. Sandro was amazed at this. Like many geologists familiar with a well-studied region, he had come to believe that no important new discoveries could be made there.

My smugness at this find was short-lived; I learned my own lesson in hubris soon after. I had come to believe that the impact layer itself marked the extinction not only of swimmers in the sea, such as fish and ammonites, but also of forms such as worms and urchins, creatures that ingest sediment, pass it through their guts, and strain out any organic matter. The long movement trails or fecal trails that these organisms leave in the sediment are readily preserved over vast amounts of time. These marks, which are called trace fossils, are records of ancient behavior, and I had come to believe that a dramatic near-disappearance of these fossils just beneath the K/T boundary interval in virtually every region where I had studied this event was due to their extinction. Excitedly I told Sandro of my hypotheses and asked to see the K/T boundaries in his region. Sandro looked at me wryly. "So you think the trace fossils, so abundant in the last beds of the Cretaceous and so rare in the overlying beds, were wiped out?" Yes, I answered, and why not? If so much else from plankton to dinosaurs was killed off, why not oceanic bottom feeders as well? He laughed at me gently and took me to see the rocks.

We sampled large blocks of strata, filled with trace fossils, from levels found immediately below the thin clay layer bearing the iridium, spherules of glass, and shocked quartz that mark the great impact event. Later that night, back in his lab, Sandro cut the rocks we had gathered with a diamond saw, polished them, and showed them to me under the microscope. The trace fossils were easily visible, and each was darker than the surrounding rock. They were also filled with small green spheres. With disbelief I saw that the burrows were filled with material that could only have come from the *overlying* impact layer. Unless the worms of the Cretaceous Period had mastered the art of travel into the future, there was no way the burrows I saw could have been made before deposition of the clay layer generated by the impact. Each burrow was lined with tiny spherules and rock that had been blasted upward into space from the Yucatan crater itself, to settle back on earth as a thin sprinkling. In Italy, this fine spherule-rich material settled down through the deep ocean, finally to form the thin, iridium- and spherule-rich clay layer at the bottom of the deep Italian sea bottom we had sampled.

The implications were pretty clear: All of the trace fossils we saw were created not by creatures living before the impact but by survivors of that event, worms that burrowed into the sediment to mine the rich deposits of death: the newly killed bodies of the Cretaceous world. We were looking at the hearty feast of the few creatures in this region that *did* survive the extinction; they ate dead Cretaceous bodies, and I ate crow. Sandro had saved me from making a monumental blunder, which I had based on a preconceived notion. Long after, as I pondered another, younger extinction, I wondered how often my colleagues debating the extinction of the Ice Age mammals had fallen into a similar trap, that of knowing their subject too well to recognize the significance of new data that didn't "fit."

In many ways, the debate about how the Age of Dinosaurs came to an end is similar to the current debate about the large mammal extinction at the end of the Ice Age. In the Cretaceous–Tertiary extinction debate, only two camps exist: those who think the extinction was caused by the impact of a large asteroid, and those who don't really care what caused it—climate change, volcanic eruption, whatever—as long as it wasn't an asteroid. The debate is acrimonious, for the stakes are high. One side will go down in scientific history as having backed the wrong horse.

The debate between those who believe that the equivalent of a meteor—in other words, a sudden calamity—killed off the Ice Age mammals—in this case, not a comet but human hunting—and those who believe climate change did it does not seem as nasty. But the literature suggests that the two sides hold positions as starkly polarized as any in the K/T debate.

. . . .

The last leg of my trip was again overland, by train, and again through atypical autumn storms, before I arrived in Paris. This city had so often been the end of other journeys for me, a last stop before the long plane ride home. Yet now it represented a first step in my quest to solve an ancient murder. I walked through the city in the gray of late September, with many miles and much new information already behind me, a long trip taken even before the start of the voyage I needed now to begin. And against the cold skies of Paris, 1995, it was not hard to imagine wild animals proliferating here during the recent Ice Age. For this place too was the home of great wooly mammoths, beasts long since removed from the face of the earth.

How long has it been since the great beasts browsed the fertile plain we now know as Paris? In this most elegant of European cities, perhaps the least wild place in Europe, you can find a few faded memories of them still, forlorn ghosts now frozen in bronze. If you walk through the Tuileries, the great statuary park extending north from the Louvre, you'll find near the east entrance a large brass rhino in the act of goring a lion to death. Across the river, just outside the Musée d'Orsay, stands another rhino even more impressive than the first. Farther south, in the Jardin des Plantes, a mastodon stands guard over an old museum—a giant boneyard where Georges Cuvier puzzled out the nature of the vertebrate skeleton and then used his powerful new methodology of comparing similar bones from dissimilar creatures to arrive at revolutionary theories about the relationships among animals and their deaths. In Cuvier's time, during and soon after the French Revolution, when the world was so full of death, perhaps it was understandable for a young naturalist to ponder extinction so deeply, even with the very limited fossil material then at hand.

Paleontology is an odd science. We are left with the barest traces of what an extinct species was like: a few bones, part of a shell, or a trail in eons-old mud. Gone forever are the colors and sounds, the preening or predation or parental care, the food supplies and behaviors that mark each species as unique. As I strolled through the Musée d'Orsay on a quiet Saturday afternoon, the great French Impressionist art collection made me think about the similarity between how we perceive a painting and how we perceive the distant past. Walking by a Seurat, Monet, or Degas, we see wondrous images. Yet as we come closer to these paintings, the various strokes and dots emerge as separate marks, and the image dissolves into isolated components devoid of meaning. Our minds are what connect the dots and make the image.

The fossil record is a bit like that. The scarps of bone or shell, always incomplete, are woven into a picture that is part real, part imagination. We construct ancient ecosystems by stepping back from the fragmentary bones and filling the picture in by supplying—and in some cases imagining—the necessary information. Just as a Monet conveys a rich world on the basis of fragmentary information, we can conjure up the past without a time machine, and even without complete skeletons. But what of those paintings where so little information is present that no recognizable picture emerges? That was the situation the pioneering paleontologists faced as they dealt with tiny scraps of bone, or perhaps large skeletons that matched no known living creatures, at a time when all animals that had ever existed were thought to exist still.

Cuvier saw right through the romanticism of his contemporaries. He destroyed the concept of the Lost World. The great Ice Age mammals were key players in the early 19th-century debate on whether anything has ever gone extinct. (Today that debate seems ludicrous in light of estimates suggesting that as many as 100 species a *day* are currently disappearing from the face of the earth.) In any case, Baron Cuvier and others finally convinced their doubting colleagues that such great creatures as the woolly mammoths and ground sloths were simply too large to be holding out in some lost corner of the earth. It was also apparent in the early 19th century that North America had lost even more of its Ice Age fauna than Europe had. As Darwin lamented in 1836, "It is impossible to reflect on the state of the American continent without astonishment. Formerly it must have swarmed with great monsters; now we find mere pygmies compared with the antecedent, allied races."

With the realization that a great bestiary had gone extinct, and relatively recently at that, scientists studying the phenomenon naturally began searching for a cause. Cuvier hypothesized catastrophic floods, while his less secular colleagues pointed out that the global flood described in the Bible would have done the job nicely as well. Louis Agassiz, discover of the Ice Age, assumed that the world had been completely covered in ice, not flood water, but that the results were the same—massive extinction. Still other naturalists looked beyond such natural catastrophes to another potential agent of destruction.

Charles Lyell is considered the father of modern geological science. His texts on the subject influenced many of his contemporaries, including Darwin himself. Thus, when Lyell suggested that human activities might have produced past extinctions, many naturalists listened. Lyell noted prophetically, "We must at once be convinced, that the annihilation of a multitude of species has already been effected, and will continue to go hereafter, in a still more rapid ratio, as the colonies of highly civilized nations spread themselves over unoccupied lands." Initially, Lyell was not convinced that humans and the Ice Age beasts were contemporaries, but as the evidence of human antiquity mounted, Lyell began to suspect that humankind had at least helped exterminate many of the Ice Age mammals. He remarked that "the growing power of man may have lent its aid as the destroying cause of many Pleistocene species." He also noted that we humans wield the sword of extinction as we advance. Could it be that 10,000 years ago that particular sword was made of fluted stone spearpoints?

Cuvier believed that the world of the Ice Age mammals had ended in catastrophe. He had ample reason for this judgment. Having seen, in rock outcrops in the region near Paris, how abruptly the Cretaceous Period seemed to have ended, he came to believe that the Age of Dinosaurs had been brought to a close by some catastrophic event. He was among the first to recognize this. He and I have walked the same French outcrops and seen the same strata, and I have come to the same conclusion that Cuvier held to his dying day: Mass extinctions, whether among dinosaurs or Ice Age mammoths, can *only* occur through sudden, rare, and highly unlikely catastrophes.

Cuvier was sure that life on Earth was punctuated by rare but enormously important "revolutions." He doubted that the cause of these revolutions would be easy to determine but noted that discovering the nature of these catastrophic events that have shaped the history of life on this planet is "the most important geological problem." Nearly two centuries later, most paleontologists would agree.

Triceratops skull. (Harvard University. Photo by Karin Hoving.)

3

When Worlds Collide

We were at opposite ends of the room and on opposite sides of the debate. We talked at each other, and the words simply passed by, heard but dismissed. In the front of the room, at the podium, Peter Sheehan and I were facing a hostile crowd. We were at that moment the target of the scathing sarcasm and booming voice of that most overpowering public paleontological presence, Dr. Robert Bakker. There is no scorn to match Bakker's, especially when he is performing before his adoring faithful. Yes, Peter Sheehan and I were being excoriated by Bakker for the very reason Stephen Jay Gould wrote about us in his recent book, *Dinosaur in a Haystack.* We both believe, and have shown through paleontological observation and experiment, that the dinosaurs died out in sudden catastrophe, and Bakker believes anything but that. At the moment he was lambasting us about salamanders. "If your meteor hit," he bellowed, "how could salamanders have survived? And where are the fried dinosaur bones?" The rest of the audience nodded sagely, if in some confusion. Salamanders, yes, good point, Bob.

I was again attending a large scientific meeting, yet this one was as much a public relations extravaganza and public spectacle as a legitimate scientific colloquium. *Dinofest 2* was a month-long exhibition of dinosaur displays and skeletons assembled at the Arizona State University in Tempe. By the time I arrived, tens of thousands of kids and parents had already walked among the assembled dinosaur skeletons—and mammoth and mastodon skeletons as well—in a large pavilion. To finish the long festival and give it an aura of scientific respectability, about 50 specialists on dinosaurs and their world had been brought to Tempe for four days of symposia and talks.

All the big-name dinosaur guys were there, and the two biggest of all, Jack Horner and Bob Bakker, could easily be found simply by looking for the biggest crowd. As each passed through a room or hall, a retinue of attendees, groupies, and curious onlookers followed.

The conference itself was wonderful. Where else could you hear scientists argue passionately about creatures so long dead? Every aspect of dinosaur life was invoked, from breeding and egg laying to feeding and predation. For four days the dinosaurs were brought back to life. And then, on the last afternoon of the conference, a few killjoys were asked to kill them off yet again and to explain to the faithful *why* there are no longer any dinosaurs. We poor Mesozoic undertakers were surely the most unpopular people at the conference, not only because we had to bury the great icons but also because of the way at least some of us say they died.

It is a mystery to me why the dinosaur faithful are in denial about the extinction of the scaly behemoths. Bob! Jack! They're dead! Move on! Get over it! Jack Horner gets cranky whenever the topic comes up and repeats that he would much rather think about the dinosaurs' life than their death. *That* solves the problem! Bob Bakker thinks that wandering dinosaurs brought diseases to the various dinosaurian legions and that they died in a manner reminiscent of Hollywood's *Outbreak* as pestilence spread. Virtually all dinosaurs specialists seem to believe that the dinosaurs were in decline well before their final extinction and that the mass extinction at the end of the Cretaceous was a slow dance of death and mammalian replacement. Other than Dale Russell, there are no catastrophists—those who believe the comet did it—among the dinosaur faithful.

Two themes emerged among those trying to explain away the end of the dinosaurs. First, if the dinosaurs were wiped out by a meteor, they argue, why is it that they were in long decline prior to this supposed catastrophe? And if they were in decline, then surely their end had nothing to do with a sudden impact. Second, if they were suddenly killed off by the blast effects of a meteor, where are the bodies? Such a catastrophe, the argument goes, would surely leave great piles of dinosaur bodies and bones.

I have heard these two arguments for a decade now, and I have constantly wondered at their logic. The arguments appear at first glance to be based in common sense. It does seem that the agent or agents causing a long decline would be something other than a sudden catastrophe. And, it would seem logical that a sudden rapid catastrophe would leave behind piles of bodies—or, in this case, bone beds. Yet both of these premises are false.

A sudden catastrophe could as well occur at the end of a long decline, caused by some other agent of change, as occur after a period of long population expansion. That is the point. Sudden catastrophe is brought about by some new factor. It arrives independently of normal conditions. And Peter Sheehan (and I) even dispute the hypothesis that dinosaurs were in slow decline.

Second, there would not have been a bone bed even if every dinosaur had died out in a matter of days or weeks (which Sheehan and I both believe). If every human on Earth died *today*, it might seem that all those bodies would create, in the fossil record, some sedimentary layer filled with human bones. But in reality, *every* human on earth *today* will be dead sometime in the next hundred years, and the amount of sediment laid down in any year is virtually the same as that deposited in any century. A sudden catastrophe therefore looks no different from the normal life-and-death processes. Most of the dinosaur specialists among my colleagues cannot—or will not—understand this point.

Just as there is not, and cannot be, a dinosaur bone bed marking the end of the Cretaceous, there cannot be a mammal bone bed marking the end of the Ice Age. Yet those two arguments—they were dying out beforehand, and we can't find the bodies—are the major arguments put forward by Ice Age specialists to refute the possibility that the large Pleistocene mammals died out catastrophically and suddenly. If humans wiped out all mastodons and mammoths in North and South America in 2000 years, such a destruction would leave virtually no trace in the record.

I met Bob Bakker for breakfast the next morning. We discussed the extinction issue (much more cordially this time, being in private) and went on our way. But our conversation, and the entire conference, convinced me even more that to understand the passing of the mammoths, one must first understand the passing of the dinosaurs. First, because if the dinosaurs had not gone suddenly extinct as a result of the most improbable of chance events, then mammoths and humans would probably never have evolved. This is perhaps the most wondrous aspect about the mass extinctions. They *provoke* new evolution. The Age of Mammals did not just come about. It happened because the world was emptied of larger creatures 65 million years ago, favoring the evolution of new forms. Second, because from any individual, sudden mass extinction, we can learn many lessons that apply to all such events. As Cuvier believed, we must know *all* of "life's revolutions" to know any one of them.

. . . .

Let us re-enter our time machine and travel back to the last day of the Mesozoic Era, a real day that occurred some 65 million years ago, give or take a million years. It is unfortunate that we cannot pin down the actual decade, century, or even millennium, let alone the specific day; such imprecision for an event of this magnitude seems unfair. The end of a world deserves a more precise date. But our methods of age-dating are too imprecise to come any closer, for all of our great isotopic machines and laboratory techniques can at best give a plus or minus in terms of hundreds of thousands of years. What we *can* do, however, is make reasonably precise estimates about what happened in the days, months, and years following the impact of a gigantic comet or asteroid in what is now the Yucatan Peninsula of Mexico. The enormous crater resulting from that event is known as the Chicxulub structure. Its formation transformed the world, and ended the Age of Dinosaurs.

What would the last dawn of the Mesozoic have been like? Probably, like so many shattering events that shape our own lives, it started in unexceptional fashion. How many people wake up knowing it is the last day of a life, or that they are seeing a loved one for the last time? I believe the last Mesozoic dawn was utterly ordinary for the doomed creatures inhabiting that lost dinosaurian world.

Would the last day have been set against a backdrop of exploding volcanoes? As a child I loved the wonderful stop-action dinosaur movies. In most of these movies, the end of the dinosaurs was neatly proscribed: They and their world, island, plateau, or whatever were consumed by bubbling, burning lava and by thunderous, crust-ripping earthquakes. Any dinosaurs that didn't fall into the lava fell into cracks within the earth. Clear symbolism: Hollywood voted for a volcanic end to the Age of Dinosaurs. Until recently, however, paleontologists have had a much different view. They saw the last days of the dinosaurs as the fall of a once-mighty empire, before the pitiless onslaught of those late Mesozoic Visigoths, the mammals. According to this view, it was the competitive superiority of egg-eating mammals, or perhaps the general virility of the new mammalian overlords, that caused the ancient race of dinosaurs to fall to dust at the feet of its furry-footed conquerors. In this scenario we can imagine the last dinosaur dying in a manner as ignoble as the last Dodo: unloved, unmourned, unremarked.

Although a few diehards (most notably the dinosaur experts themselves) still cling to the old view, most specialists in the field now see things differently. The last day wouldn't be marked by the death of some last single dinosaur, perhaps rotting with old age and hopelessly searching for a mate that is already gone. It would not have dawned on a world already populated by the new overlords, the mammals surrounding our last dinosaur like greedy relatives awaiting their inheritance, or on a world where great volcanoes spewed forth death and brimstone. The last day of the Mesozoic surely started in an unexceptional way. But it ended in one of the unique events of our planet's long history.

On the basis of our reading of the fossil record, we can envision some fragments of what the last of the Mesozoic world was like. In North America, the great inland sea that had dominated the geography of the continent for more than 50 million years had long since dried and been replaced by large inland lakes and swamps. In what is now Montana and the Dakotas, slow rivers carried sediment east and south from the eroding Rockies, much as they do today. It was a world of life, with a rich diversity of trees and flowering plants, themselves at that time a relatively recent evolutionary innovation gradually replacing the once-dominant conifers. There must have been a feeling of both newness and great antiquity to this world, as both the ancient race of reptilian dinosaurs and the emergent insects and smaller animals (including our ancestors, the Late Cretaceous mammals) evolved in tandem with this new world of flowering plants.

In some ways there surely would have been a feel of Africa as well, or at least of Africa as I picture it: a gently rolling landscape with numerous large animals visible on all sides; thick, sluggish, sediment-choked rivers roiled by the scaly backs of giant crocodiles; great flocks of birds; and demons that looked at first like birds but on closer inspection were revealed to be hideous, gargoyle-like apparitions, pterodactyls and pteranodons that must have ruled the skies. There was no grass in this world, but the Late Mesozoic river valleys would have been even more lush and green than the baked yellow grassland that is characteristic of Africa today. The thick, humid air may have felt quite different, for we have determined that in the Late Cretaceous, the oxygen content of the air was up to 10% higher than in the modern world. Forest fires would have been more common, widespread, and catastrophic because of the higher atmospheric oxygen; perhaps they were the greatest hazard facing most of the creatures of this world. It would have been a time of beauty, a time of wonder, a time of life.

Dinosaurs, of course, would have been the most obvious actors on this stage. Great lumbering *Triceratops*, herds of duckbills, the rapacious tyrannosaurs, a few small sauropods, bevies of smaller dinosaurs scurrying about on their hind legs in search of food. Looking perhaps like antelope or wildebeest or great herds of rhinos, but acting more like birds, these great herbivores would have mingled and browsed near the rivers, nesting, fighting, preening, displaying. Although our traditional view of dinosaurs is of the old and clumsy, the stupid and cold-blooded, a great revolution in understanding has led us these last three decades of the 20th century to view them as far more intelligent and active. It was once far easier to rationalize their final passing as some normal event in the natural order of things: the obsolescent giving way as the intelligent, furry, warm-blooded mammals with superior parenting skills took over. But in our new view, dinosaurs seem far more modern, and the nature of their passing becomes more mysterious. That they ended with a bang rather than a whimper seems somehow fitting.

There is now little doubt that the bang that ended their world was impressive indeed. Did any of the creatures take note of the dim celestial light growing brighter each night as the giant comet raced inward toward the sun from its deep-space birthplace? Was its ever-larger tail a distraction to the night-flying fauna of the latest Mesozoic, the insects and newly evolved birds, or the soaring pterosaurs and other saurians of the Late Cretaceous? Did the head of the comet eclipse the moon in brightness as it hurtled inward, those last few nights of the dinosaurs? In those last days, the comet plunged sunward—and coincidentally earthward—at 25 kilometers a second, 90,000 kilometers each hour, passing inward across the moon's orbit in its final few hours, traversing the distance from the moon to Earth in a bare 4 hours.

Texas was then, as now, a flat plain, but it was far wetter and more fertile than it is today. From the fossils left on its chalky shores, we can reconstitute ancient Texas as part land and part wide, shallow sea, each teeming with life. The wide lowland would have been swampy and vegetated, a jungle of broadleafed flowering plants and soaring conifers, its sea warm and clear. The subtropical environment, then as now, was packed with life, an assemblage we would find both foreign and familiar. Great saurians languidly splayed in the mud of slowly moving rivers, and the screeching cacophony of birds and bird-like things would perhaps seem modern. But the faster-moving dinosaurs both large and small would surely seem surreal. Mammals were not dominant or even common. Like the rats of our world, they inhabited the cracks and corners of this one: They were night creatures, cave creatures, high-tree dwellers. Rat-sized to cat-sized, they would seem poor cousins, much like the opossums of today: otherworldly, slow, and stupid. The mammals of that time were surely nocturnal, judging from the large eye sockets in the few spectral skulls preserved from those last days of the Mesozoic world. And who could blame them, given the absolute dominance of the larger, more ferocious, and perhaps smarter dinosaurs? By the late Cretaceous the dinosaurs had dominated the world for over 160 million years. During all that time, a period incomprehensible to us short-lived humans, dinosaurs and mammals coexisted. And in all of that time, the mammals had never even come close to the dinosaurs in size, diversity, biomass, or ecological importance.

. . . .

The Chicxulub comet was unexceptional as comets go; it was yet another bit of the primordial solar system, a hunk of rock and ice perhaps left over from the origin of the solar system long billions of years ago. For more than 4 billion years it had hung well beyond the orbit of Pluto, among 20 billion other hunks of dusty ice in orbit far from the bright star controlling its fate. Yet after such unknowable time, its long, slow orbit underwent more radical changes, as the infinitesimal tugs of gravity from the other denizens of the solar system exerted their tiny but ultimately fateful effect. Eventually it began a new journey, a long fall toward the sun. It crossed the orbits of the planets. Perhaps it led a life of many such orbits, falling inward toward the sun, crossing the outer, then the inner planets. Flashing across the orbits of Mars, of Earth, and then of Venus, its water and other frozen gases boiled to life and shot back into space, creating a great tail streaming outward away from the sun. As it flashed around the sun at great speed, it was thrown outward, like a rock in a sling, back into reaches of cold space, slowing now, heading toward deep space until once again it balanced between the call of the now-distant sun and that of the vastly more distant stars. It rested a moment in Newtonian grace and then once again began to slide inward. How many times did this particular comet take this long ride, growing smaller with each journey inward as more of its surface boiled away in the radiation-rich environment of the inner solar system?

Or was this comet of a different sort? Was it a far larger piece of cosmic flotsam, more massive than the other denizens of the nearly deep-space Oort cloud, home of comets great and small, the last frontier of our solar system well beyond Pluto? Was it 100 kilometers in diameter, filled with iron and sulfur, rich in heavy elements, composed of atoms that were originally born in the crucible of some far supernova soon after the birth of the universe itself some 12 to 18 billion years ago and finally coalesced in the neighborhood of our own star, Sol? And was this hunk of rock and ice an eventual slave to the other great source of gravity in our solar system, Jupiter? This may be a far more likely scenario. Falling inward, the great asteroid would have been captured by Jupiter's immense gravitational field on one of its inward journeys and pulled into smaller pieces as it sped in complex cartwheels around the Jovian system, much of its mass ultimately crashing into the roiling atmosphere of the great planet. Gene Shoemaker, codiscoverer of the comet Shoemaker-Levy 9, which smashed into Jupiter a few years ago, suspects that Jovian comets are often flung back into space after being captured by Jupiter. They are torn apart, and most fragments are ultimately destroyed in the lightning-quick fall into Jupiter itself. But some chunks may be whipped *back* into space, shot from the gravitational cannon of Jupiter, taking accidental aim at any targets sunward, first crossing the orbit of Mars, then of Earth, then of Venus and Mercury as they speed toward the sun.

Whether as a smaller comet passing sunward or as a fragment of a Jupiter-perturbed body, the Chicxulub object won the lottery of destruction. Imagine betting a dollar in a state-run lottery and winning the million-dollar prize—and then taking a dollar of your jackpot and winning the prize the following week as well. And yet the odds of this would be far better than those of the Chicxulub object crossing the Earth orbit at exactly the moment that Earth was in its path. Nevertheless.

In 1996 I gave a lecture in Duluth, Minnesota, about the collision of the Chicxulub object with Earth. It was early February, and my lecture coincided with some of the lowest temperatures ever recorded in the United States. It was a perfect backdrop for a lecture about the end of the world. At the end of the lecture a man asked me, "With such great odds against such an event happening, aren't you now convinced of the existence of God?" This question gave me pause. Such unbelievable coincidences certainly make for inferior fiction. What *are* we to make of them when they are true? The best I could answer was that those dinosaurs so long ago had a really bad day.

Large meteors have hit our planet throughout the history of the solar system. In the early days of our solar system, meteor or comet impact was the norm, a daily blitz of Wagnerian proportions. But with passing time came a lessening of the stony rain. Pieces coalesced with already-formed planets, were swept into the sun, or maintained a brooding existence in the cold reaches between Mars and Jupiter or even farther away, in the Oort cloud, well beyond Pluto. By the time life first appeared on our planet, more than 2.5 billion year's ago, the rate of impact must have perceptibly lessened, and it has continued to do so since. Even so, our planet has been hit from time to time, and occasionally during the last 500 million years, by significantly sized objects. In the Triassic Period, some 215 million years ago, a great asteroid or comet smashed Earth and created the Manicouagon Crater, now a lake in Quebec about 100 kilometers in diameter. Creatures died following this event, but not in numbers sufficient to reconfigure the biological structure of the earth. Later in the Mesozoic, other great collisions occurred, creating several craters of 60 kilometers or more in diameter. Yet even these giant craters are but a fraction of the size of the hole left by the Chicxulub object; nothing on Earth (or Venus and Mars, for that matter) now recognizable as a crater can match the 300-kilometer diameter of the Chicxulub crater, a hole so large that it was not even recognized as *being* a crater until the late 1980s.

What would it have meant for the world (and for the history of life) if that long-ago comet had been but minutes earlier or later? What if there had been the nearest of misses, a cosmic lightshow turning even the most phlegmatic saurian heads, changing breeding patterns perhaps, creating great storms of a roiled atmosphere by the near miss? What-ifs are so useless but so much fun. What if Christ or Mohammed had not been born, or Hitler or Stalin? What if the American fleet had lost its carriers at Pearl Harbor or if Napoleon had succeeded in conquering Britain? Would our world today be anywhere near its current political or economic composition? Doubtful. Very improbable. In analogous fashion, what if the asteroid had not hit, saving the dinosaurs from extinction? Would there be an Age of Mammals? Would evolution have proceeded on lines resulting in (along with much else) the rise of humans, or any other consciousness? I think not. Perhaps the dinosaurs would be extinct by now, perhaps not. But the rapid expansion of mammalian evolution in the ecological vacuum following the great K/T mass extinction would surely not have taken place. We owe our existence to that hammer-blow from space.

. . . .

On that long-ago Texas plain, then partially land and the rest shallow sea, how many creatures watched the fiery end of their world?

Ground zero was a shallow sea: warm, coralline, perhaps 10 meters deep. Creatures familiar to us, such as oysters, clams, and shoals of silvery fish, were abundant. But less familiar creatures inhabited these clear, sunlit seas as well: ammonites, passing over the bottom in shelled splendor; archaic squid; and, perhaps most impressive of all, the giant reptiles such as mosasaurs and plesiosaurs. The surface of the sea was alive with a rich tide of plankton. The bottom of this wide, shallow sea was made up of limy

Marine reptiles from Age of Dinosaurs. (From L. Figuier, La Terre Avant Le Déluge, 1864.)

carbonates, but deeper down were sediments deposited when the ocean had been large saline lakes, producing thicknesses of salt and sulfate-rich gypsum resting just beneath the muddy sea floor.

The Chicxulub object hit this shallow sea traveling at least 25 kilometers per second. At such speed, Earth's atmosphere provided no cushion or brake; the incoming celestial body vaporized all gas molecules in its path, leaving a short-lived column of vacuum. With the impact, enormous energy was released as light and heat. For hundreds of miles around, every living creature must have been instantly destroyed.

The object, which was made of rock, water, and metal, disintegrated from the force of the impact. It destroyed Earth's crust in the region to a depth of at least 5 kilometers, creating a crater 300 kilometers across. A volume of Earth's surface many thousands of times larger than the incoming mass was blasted upward into the atmosphere and into space. Some of this material followed a suborbital trajectory, rising tens of miles into the low atmosphere, eventually to fall back to Earth. Much more went higher still, ultimately attaining a low Earth orbit. Some escaped Earth's gravitational field altogether. Most, however, came back to Earth fast, hard, and hot. A great deal of elemental sulfur, until that moment locked safely in Earth's shallow strata, was released by the blast to wreak havoc with the animals and plants on the unlucky planet in the weeks and years ahead.

. . . .

There are odd ironies in the concept of extinction. In our current world, hardly a day goes by without some media reference to the Endangered Species Act, to some new outrage of environmental degradation, or to estimates of extinction rates themselves. Yet the fossil record suggests that *producing* extinction in any individual species can be difficult. For a species to go extinct, every individual must die. For some species, such as the narrowly adapted (those that have very narrow tolerances for temperature, for instance, or require a very specific food supply), removing entire populations through death is no onerous task. Some species of Hawaiian land snails are entirely confined to a single tree; chop down the tree and you have produced an extinction. But what of the more widely distributed species, such as swimmers and flyers that inhabit large areas of the earth's surface? In this case, exterminating every single individual requires extraordinary conditions.

Chicago paleontologist David Raup has estimated that through much of geological time, a species disappeared every several years or centuries, a far cry from the number currently estimated to be going extinct. Yet during the episodes of mass extinction—those short intervals of intense species death that have periodically occurred in the earth's history—the extinction rate has gone even higher. Such was the case in the aftermath of the Chicxulub impact. All evidence seems to indicate that this brief catastrophe was the most intense interval of extinction that has ever occurred.

One of the most perceptive comments that I have heard about the causes of mass extinction came from Buck Sharpton, discoverer of the size of the Chicxulub crater. Buck maintains that mass extinctions are caused by changes in the global atmosphere. These atmospheric gas changes, such as a change in the volume of some particular constituent of the atmosphere, can be produced by many things: asteroid or comet impact, volcanoes, a change in sea level. But Buck's view is that the actual lethal agents that kill off enough animals and plants to produce a mass extinction are changes in the makeup and behavior of the atmosphere or in factors (such as temperature change and deviations in circulation patterns) that are dictated by properties of the atmosphere. At least for the K/T extinction, those few scientists who worry about killing mechanisms tend to agree. After all, the Chicxulub object itself merely squashed a few mollusks at ground zero. Its effect on the composition of the atmosphere, however, was far more lethal.

Just *how* lethal was recently illustrated in two studies conducted by NASA scientists. In the first, a computer model estimated that between 0.4 and 7.0×10^{17} grams of sulfur (or about 10 to 100 billion tons) was released into the atmosphere after the K/T impact and then fell back to earth during the weeks and months that followed. A small portion of this sulfur fell back as acid rain, polluting the lakes and streams of the land areas and the upper few meters of the sea. Although this acidification may have been a killing mechanism, it was probably more important as an agent of cooling than of direct killing. However, more deleterious to the biosphere may have been the reduction (by as much as 20% for 8 to 13 years) in the solar energy reaching Earth's surface. According to most sources, this reduction would have been sufficient to produce a decade of freezing or near-freezing temperatures in a world that had been largely tropical. The prolonged "impact winter" (Carl Sagan's term for the aftereffects of both atomic war and asteroid impact) is thus the most important killing mechanism, and it was brought about by a vast increase of aerosols in the atmosphere over a short time.

Another model examined the global climatic effects of atmospheric dust produced by the impact of a large (10-kilometer) asteroid or comet. According to this model, fine dust generated by the impact produced a blackout of the sun for several months. Even in the tropics, the world awakened to a dim twilight at best—light levels too low for photosynthesis. Thus plant life dwindled. But perhaps the most ominous change occurred in

Earth's water cycle. Computer models suggest that in the post-Chicxulub world, global precipitation was reduced by more than 90% for the first several months after the impact and was still only about half normal by the end of the first year. The world got cold, dark, and dry. This is an excellent recipe for mass extinction, especially for plants—and the creatures that feed on them.

Those were the "long term" effects. Much more harrowing would have been the days after the impact, when much of the world's forests caught fire, producing the single largest forest fire in the history of this planet.

. . . .

The herd of duckbills lowed to one another as the first hints of dawn began to paint the eastern sky with the faint promise of daybreak. The low coastal swampland was already alive with the calls of birds, but calls somehow dissonant and anxious, as if predators were approaching, though none could be seen or smelled. To the west, still black with night, the full moon was beginning to set; to the south, however, a bright star blazed forth, with a long phosphorescent tail extending across half the night sky. It had been in the sky for many nights, growing larger and brighter with each reappearance. Among the vertebrate and invertebrate fliers that depended on the moon for navigation, this new beacon scrambled all the visual neurons; nowhere in the intricately coded DNA of the birds, insects, pterosaurs, and pterodactyls was there any information about a second moon in the sky; it troubled them at a deep, intracellular level. Seaward, in the warm reefal ocean, the combination of approaching day, a full moon, and the new celestial light meant extra food for the piscine swimmers and shelled ammonite predators as they hunted crabs and shrimp in the coralline cities.

The dinosaur herd was completely awake now and starting to feed on the rich low vegetation. A pink glow of dawn was accompanied by swarms of insects rising, many settling on the browsing dinosaurs, others flitting through the humid, oxygen-rich atmosphere. Overhead, the tail of the great comet began to disappear as the night sky was overtaken by dawn, but its disappearance was as much a matter of its movement now as of its being overpowered by the approaching day. The bright head of the comet could be seen to descend slowly into the south and finally disappear below the horizon, to be followed some seconds later by the orange glow of a second dawn.

From the southern horizon a thin but brilliant bar of white light shot upward into the sky, the first proclamation of the end of an era and the beginning of a new one. Molten rock from the impact created this beam of light as rock from the comet and the impact site were intermixed and blasted into the thin pillar of vacuum created by the comet's fall to earth. The dinosaur herd was still oblivious; southern lightshows were irrelevant to brains programmed to seek food, avoid predators, and propagate. But the great flocks of birds paid heed, falling silent as the second dawn unfolded. The thin pencil of light began to change color, become more diffuse, and widen; from its base, tiny specks of light fanned outward in all directions. What sounded like distant thunder now silenced the dinosaurs as well, and all turned to the south as a bass rumble intensified. A low cloud appeared in the south, moving rapidly toward the herd. The dinosaurs turned in fear as the shock wave approached and then passed in speeding seismic fury, emptying the trees of birds, creating great flocks of screaming avian and reptilian flyers in the rapidly brightening sky. The now-terrified dinosaurs rushed in all directions, oblivious to the silence that once more filled the landscape, oblivious as well to the orange cloud creeping upward from the southern horizon.

The first of the meteors streaked overhead, buzzing like a mad hornet emerging from the direction of the orange cancer devouring the sky from the south. A second meteor came screaming inward, hitting the sea some miles from shore with a loud explosion. Others passed overhead or fell earthward with increasing frequency, and still the sky filled with shooting stars. The sun, rising at last in the east but no match for the fireworks overhead, was soon obscured by the rising walls of smoke, for the meteors were now crashing landward with promethean fury.

The superheated bits of rock falling from the sky began to set the great

Mesozoic rain forests alight. The fires were isolated at first, contained by the wetness of the coastal plain. Gradually, however, they began to link up, as the air and forests heated and as more and more shooting starts streaked inward. The sky overhead was now brilliant yellow, a daylight of shooting stars, with greater and lesser hunks of the Yucatan sea bottom and celestial debris commingled into artillery shells. The forests now burned in general conflagration, fueled by the thick wood and abetted by the richly oxygenated atmosphere whipped into huge winds. Within the first few hours, the vegetated regions were sterilized, as unnumbered creatures great and small were exterminated by the fire and heavenly brimstone, for this close to ground zero was Armageddon.

Death also began to spread its net in the nearby sea, as the shooting stars pummeled the shallows, rapidly raising the temperature of the top meters of the ocean. Those creatures capable of swimming began to move downward toward deeper, colder water, finding refuge if they were prompt. Soon, however, even the most vigorous downward swimming became irrelevant, as a monstrous current tugged all of the ocean dwellers toward the south. In this place, no depth was safe. Muddy sea bottoms as much as 100 feet below the normal surface were exposed by the rapidly retreating sea, and burrowing species became involuntary intertidal creatures, gasping or writhing in this first exposure to air.

Along the shorelines, land creatures chased the sea, fleeing from the blazing forest. And still the sea retreated, the hallmark of a tidal wave, moving away from the old shoreline faster than any animal could run, exposing ancient sea bottoms that had been covered for thousands of millennia and leaving behind dying colonies and marine organisms: a windfall for the terrestrial predators if any had been concerned about food. It must have been a world gone mad, a world topsy-turvy: the green, cool forests now aflame, the once-calm sea racing away from its shores, the sky ablaze with sheets of shooting stars, the Mesozoic landscape whipped by superheated, raging winds that bore the screams of burning creatures large and small as all the furies of Hell descended on an unsuspecting world.

Swamps and lakes and especially the sea's edge became last refuges of the great Mesozoic fauna. The survivors crowded the shorelines of the once and future seaway beneath the cosmic barrage, explosions now blasting great clots of burning forest as the meteors rained inward. And through that inferno a new noise manifested itself, a great roar coming from out to sea as a huge new mountain range appeared in the direction of the retreating ocean, a mountain range growing larger every moment, a black mountain reflecting the angry red of the burning land, a mountain now a kilometer high, a mountain formed from the angry sea, the largest wave in the history of the world. The huge tsunami rolled landward like an express train, now towering high over the dinosaurs and now past them, extinguishing in a heartbeat the remaining saurians and the forest fires alike, turning the Texas shoreline into a greater ocean—and still the angry incoming meteors bombarded the Earth. Eventually the great waves subsided, leaving behind high-water marks of death, the residue of the stone thrown into the edge of this long-ago Mesozoic pond. In Texas nothing survived. Nothing.

. . . .

Far to the north and west, in what is now Montana, a somewhat different assemblage of dinosaurs greeted the dawn. Great herds of triceratops darkened the brushy plain and low swampland by their multitudes, stalked only by the rare tyrannosaurs. Smaller predators and herbivores lived among the giants, each dinosaur a link in the intricate and ancient Cretaceous ecosystem. Here the dawn broke normally, but shooting stars soon began to fill the sky. The dinosaurs glanced up nervously as the first meteors began to fall, rarely at first, but with increasing frequency, eventually setting the plain afire in a dozen, then a hundred places and causing stampedes of screaming beasts. The clear morning sky darkened as a great black cloud arose from the south; the blackness of spreading dust tinged by lightning and spitting flaming rocks grew ever higher in the sky. By midmorning the sun was covered by the dust cloud, and with the darkness came rain, a rain that grew colder and increasingly acidic as the days passed. It scalded the skin of the survivors, ate away the protective surfaces of their eggs, and poisoned plants in the ever-deepening cold. The sun would not shine again on the Montana plain for a year, and when it did, it illuminated a skeletal landscape of tropical trees killed by the cold, of frost in a place where frost had not visited in millions of years; the sun finally rose on the bleaching bones of the Mesozoic world. Here and there life still existed (just as, far into the future, we find life on our battlegrounds and other fields of slaughter), though it was not life fit for zoos or pets or prideful exhibitions: cockroaches, beetles, weeds, and many, many rats gnawing on moldering bones.

EARLY AGE OF MAMMALS FOLLOWING THE CRETACEOUS TERTIARY EXTINCTION. (FROM L. FIGUIER, *LA TERRE AVANT LE DÉLUGE*, 1864.)

4

The Once and Future Kingdoms

A WORLD ENDED. A world began.

If any being could have observed our planet from space in the immediate aftermath of the Chicxulub collision, the normally blue-green Earth would have taken on an orange tint as the great Mesozoic rain forests burned. Days and weeks later the orange would have darkened, to be replaced by a dingy gray, and then black, as the atmosphere filled with soot and dust following the comet's impact. After some weeks the fires abated, after some months the sky began to clear, and the acid rain soaking much of the planet diminished. Most species on Earth were already dead. Some were still dying. A few others began climbing out of their holes and burrows or deep sea haunts to begin the task of living once again. And evolving. It was a time of death but also a time of opportunity. The dinosaur overlords were gone.

For many millennia after the Chicxulub impact, the world was turned upside down. The giant crater itself, a festering boil on the earth's surface, eventually filled with water and was quenched. The rest of the planet was not so easily restored to any sense of normality, or any vestige of the long, lazy jungle world that had characterized the last 50 million years of the Age of Dinosaurs. Climates were changed; warm areas turned cold; dry areas turned wet. All but the hardiest plants died, to fall into cold dead swamps. New plants arose in the place of the old Mesozoic forests, plants adapted for the colder climate—ferns at first, then larger bushes, and finally trees—and soon vegetation ran riot across the landscape once again, growing in abundance and density until the vegetational aspect of the world was vastly different from that of the Mesozoic. The jungles grew thicker, and in the absence of the great saurian herbivores, the very nature of the forests was transformed. Thousands of species of plants had been erased from the earth by the aftermath of the cometary collision. The new species of plants that evolved to take their place did so in a world very different from that of the Cretaceous. These differences were not only related to the new rainfall and temperature patterns. With the disappearance of all the dinosaurs, plants were no longer trampled and eaten, browsed and destroyed by the herbivorous dinosaurs. Dinosaurs appear to have been present in astonishing numbers, and their combined presence clearly maintained a very particular arrangement of plants. No one species of plant could dominate in such an ecosystem, nor had the forests ever thickened to the point that sunlight would no longer penetrate to the forest floor. Now, with the dinosaurs gone, the new plant communities grew denser and more impassable than they had ever been.

The few animal survivors began to blink and wander out into daylight soon after the holocaust. They crawled from holes dug deeply; they emerged after miraculous escapes in tiny untouched refuges surrounded by charred fields; they hatched from long-buried eggs or from deep hibernation, having slept through the planetary stress that had lasted for a year or more. Among the survivors were mammals.

For many millions of years, mammals had coexisted with dinosaurs. Most were very small, lived furtively in high trees or deep burrows, and were active at night, for only at night could they venture out with some sense of safety. It was probably not only the great carnivores such as *T. rex* that kept the mammals in check during the Mesozoic Era, but also the much smaller, swift, bipedal dinosaurian carnivores of perhaps chicken or ostrich size; these were the predators that feasted on the furry, warm-blooded mammals. Other reptiles—the rapid, bird-like dinosaurs—may have successfully competed with mammals for food resources such as insects and fruit. By the end of the Age of Dinosaurs there were surely far more mammalian than dinosaurian species, but even so, the dinosaurs seemed to hold the mammals at bay—until the great extinction.

Of course, not only dinosaurs died as a result of the great Chicxulub catastrophe. Vertebrate paleontologists have shown that in the Hell Creek beds of Montana, the sole place on earth where this changing of eras has been studied in any detail, the majority of mammal species met the same fate as the 15 to 20 dinosaur species (so few in number to dominate a world!). Only one out of 11 marsupial mammals survived the catastrophe, and only slightly more placental mammals, the lineage that led to the majority of mammals found on Earth today. But some did survive—in deep holes, or along riverbanks, or perhaps in small areas spared the ravages of the monstrous fires and acid rain—through sheer luck. How they survived may always be unknown, but survive they did. In doing so, these small, scared, rat-sized creatures hit the evolutionary jackpot. To these winners literally went the spoils: At first, they won a global charnel house, where piles of rotting meat littered the landscape, and dying and falling vegetation filled the creeks and rivers. But they won far more than the sudden windfall of food yielded by the great end-Cretaceous catastrophe, for there were no more mighty saurians to confine them to marginal habitats. A great evolutionary faucet was opened, to unleash a torrent of new, mammalian morphological innovation. No longer would the mammals have to cower or live at night. Perhaps most important of all, no longer would the mammals be required to remain small. For the first time since the Permian Period of a quarter-billion years before, they could grow large.

They did so in a bizarre world. In North America, the rainfall patterns shifted drastically as the great worldwide pall of dust and grit finally settled out, revealing blue skies once more. Huge areas of the continent became marshes, lakes, and rotting bogs; enormous amounts of coal began to form.

We have an excellent model for the dinosaurs' world: eastern Africa. The ecosystems of the African game parks have served as models for the Mesozoic world in several studies, and ecologists think the east African countryside might be a good starting point for imagining what the Cretaceous flora may have looked like. But the changes that followed the great Mesozoic extinction should serve as a warning as well. Many people suspect that should the large game now living in Africa, especially the elephants, be extinguished, the plant communities of Africa would change drastically, just as the Mesozoic forests changed after the extinction of dinosaurs. It was not just the effects of the devastating meteor impact that changed the world, but also the removal of the giant herbivores.

A zoo display featuring the mammalian fauna for even the first 10 million years following the great Chicxulub impact event wouldn't have to be very large. There would be no need for great outdoor lawns or large moated enclosures for the fearsome carnivores. There would be no elephant pens, giraffe paddocks, or stagnant fetid pools for the fly-covered hippo. A few small terraria would do for most of the mammals of 64 million years ago—or a few good hamster cages. Even the two most significant lineages to emerge—the primates, which ultimately created the most widely distributed species of mammal (us), and the elephants, which became the largest of terrestrial mammals—would have appeared verminous to us, something to give the cat.

A few mammals still living on Earth are direct, largely unchanged descendants of the earliest Cenozoic mammals. You can see these relics still, if you live in the night. In certain parts of the country, if you return home after a late night out, already in the twilight world between reality and sleep, you may see low, gray wraiths scurrying forward in misshapen gait. Or later still, with the rising of a last-quarter moon at 3 A.M., these furry shapes are found rummaging in the garbage or stealing the cat's food. If confronted they act strangely, almost reptilian in their lack of expression or even robotic in their behavior; they are somehow akin to us, but not like us. The opossums of the city are the only marsupials we know well, but I don't think we know them at all. They are a remnant of remote sandstone days when our forebears still lived on insects and great dinosaurs ruled the land.

The opossum's teeth tell a tale of antiquity. Unspecialized, gleaming with the needle-like structure of cusps still undifferentiated, teeth that were to become molars, grinders, and slicing tools as the early mammals differentiated and evolved. The opossums come to us from down the long road of Cenozoic history (the so-called Age of Mammals); they escaped the great catastrophe 65 million years ago. The opossums were common then, and they probably lived a life not so different from the one they live today, living at night, living in fear, living without defense other than the ability to breed often and quickly to replace the dinosaurs that suddenly began dying in unbelievable numbers.

The death of the dinosaurs unleashed an evolutionary torrent. Freed from evolutionary constraints, the surviving mammals became a river of new forms. Within 10 million years the world was populated by thousands of new mammalian species; within 20 million years there were whales, bats, giant herbivores, and carnivores. Yet all derived from the small night creatures, the rat-sized insectivores, and the cringing marsupials.

The turning point in mammalian evolution, when the modern-day fauna finally displaced the archaic assemblages of the early Cenozoic (65 to 40 million years ago), coincided with the cooling of the earth 40 million years ago and the spread of grasslands. At this time the ancestors of today's dominant mammalian herbivores split into two groups, the odd-toed forms (which include horses, tapirs, and rhinos), and the even-toed group (comprising pigs, hippos, cattle, deer, giraffes, camels, and antelope). Modern carnivores appeared as well, diverging into the feliforms (cats, hyaenas, and mongooses) and caniforms (dogs, bears, raccoons, weasels, and seals). By this time primitive whales and other marine mammals had already invaded the seas, and bats had begun competing with the birds for mastery of the skies. But all of these new mammals were originally adapted to a world much like the one known to the dinosaurs, a warm world of steamy swamps and humid tropical jungles. By about 45 million years ago, the warm, continent-covering jungles began to recede from many lands as the earth cooled. The mammals' first great burst of evolution, following the death of the dinosaurs, was in response to a suddenly empty world. The second great burst, some 40 million years ago, occurred when grasslands and low herbs began to replace the trees. Increasingly, the creatures of our world took on a modern appearance, and all the while the earth continued to cool.

By 5 million years ago the world had markedly cooled (a consequence of the spreading apart of the continents) and dried; great expanses of tropical forest retreated toward the equator and were replaced by vast regions of grass. Savannas gave way to true grasslands and steppes, while farther north the first appearance of Arctic vegetation and permafrost signaled the ever-tighter grip of winter. As the climate cooled and the vegetation changed, mammalian species adapted.

The creation of distinct periods of summer and winter took its toll. With the advent of more pronounced seasons, the herbivores that flourished were those either better equipped to survive long periods with little food or able to make long migrations in search of seasonal food resources. Those unable to withstand drought, cold, or periods with little food went extinct. These changes all favored the evolution of ever-larger-bodied herbivores. From this cauldron the Ice Age megamammals emerged. And from it humanity emerged as well.

. . . .

Purgatorius. Is the name based on Purgatory, that place between heaven and hell where the souls of those who have died in grace must expiate their sins? Brother Webster also defines *purgatory* as a condition of suffering or remorse. Some scientific joker surely dreamed up this name and gave it to the oldest primate for which we have fossil evidence in rocks deposited soon after the end of the Age of Dinosaurs. It was about the size of a squirrel and had a long, bushy tail. If it were somehow brought back to life, who among us would recognize that we were looking at the oldest of our group, the first member of the mammalian order that eventually came to include humanity?

The earliest primates, our first ancestors, may have existed prior to the end of the Age of Dinosaurs, and the discovery of the fossil genus *Purgatorius* in the tan and black coal beds of Montana shows that they existed soon after the dinosaurs' extinction. These creatures looked much like shrews and probably acted in rather the same way; they lived in trees, and their dentition suggests a diet of insects. Between 65 and 50 million years ago, many of these early primate species were spread across the world, and all showed a variety of characteristics that distinguished them from other mammals. Some of these features, such as grasping hands and feet and mobile shoulder joints, are clearly adaptations that facilitated living in trees. The head, with its flat face, acute and forward-facing eyes (a prerequisite for binocular vision), and a relatively large brain, also may have evolved in response to an arboreal lifestyle. Natural selection acts quickly when an organism's life depends on first seeing and then *catching* branches as it swings through the trees, high above the forest floor where one mistake can be fatal. Primates show increased parental care compared with most other mammals, and they devote a long period to raising the young; a consequence of this behavior is a very low birthrate involving but one or two young per pregnancy. Small litter size may also have been related to the dangers of living high in the trees, for the young had to be carefully watched until they too could master the skills of a high-wire lifestyle.

For the first 10 million years of primate history, most of our ancient ancestors looked much like modern-day tarsiers or lemurs. About 40 million years ago, however, a new group arose: the monkeys, a lineage adapted, then as now, to life in jungles and trees. As the world cooled and forests increasingly gave way to grasslands, the primates had to adapt or disappear. They did disappear from North America, becoming largely restricted to equatorial regions, the last bastions of the tropical jungles.

By about 20 million years ago the first apes had evolved. Paradoxically, although this group is the most intensively studied of all mammalian

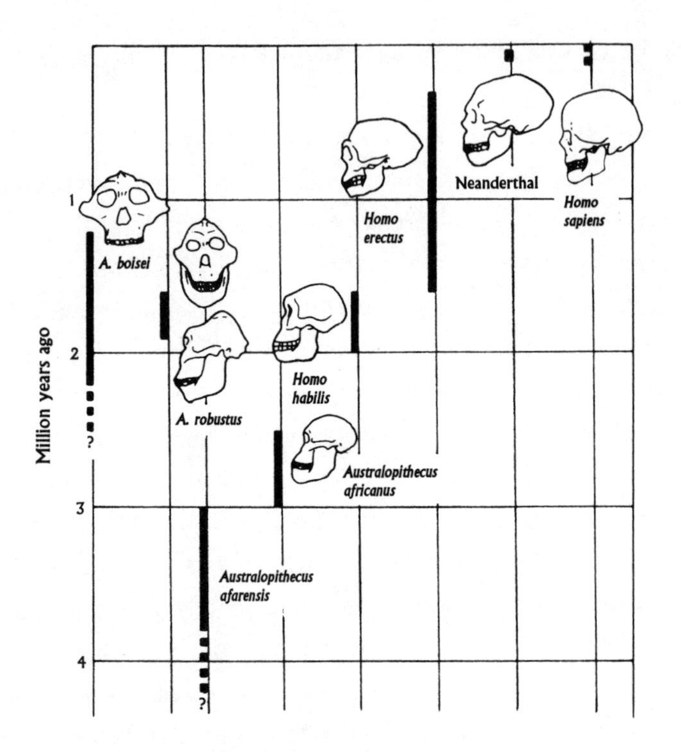

Human evolutionary changes. (Modified from Steve Stanley, Earth and Life Through Time; by permission of WH Freeman.)

taxa, the nature and evolutionary interrelationships of the Hominoidea (apes and humans) are still highly problematic. This confusion arises in no small measure from the very incomplete and spotty fossil record of apes and humans. Our skeletons and those of our ancestors didn't routinely enter the fossil record.

The earliest of the apes is named *Proconsul.* A fruit-eating primate about the size of a baboon, it shows adaptations and skeletal elements suggesting that it, like most primates, lived in trees. It certainly walked on four legs if ever it visited the ground, and it may be the ancestor of all subsequent apes and humans. *Proconsul* gave rise to a great variety of forms, and with the continental collision of Africa with Eurasia about 18 million years ago, these creatures soon spread from their African birthplace to Asia. From

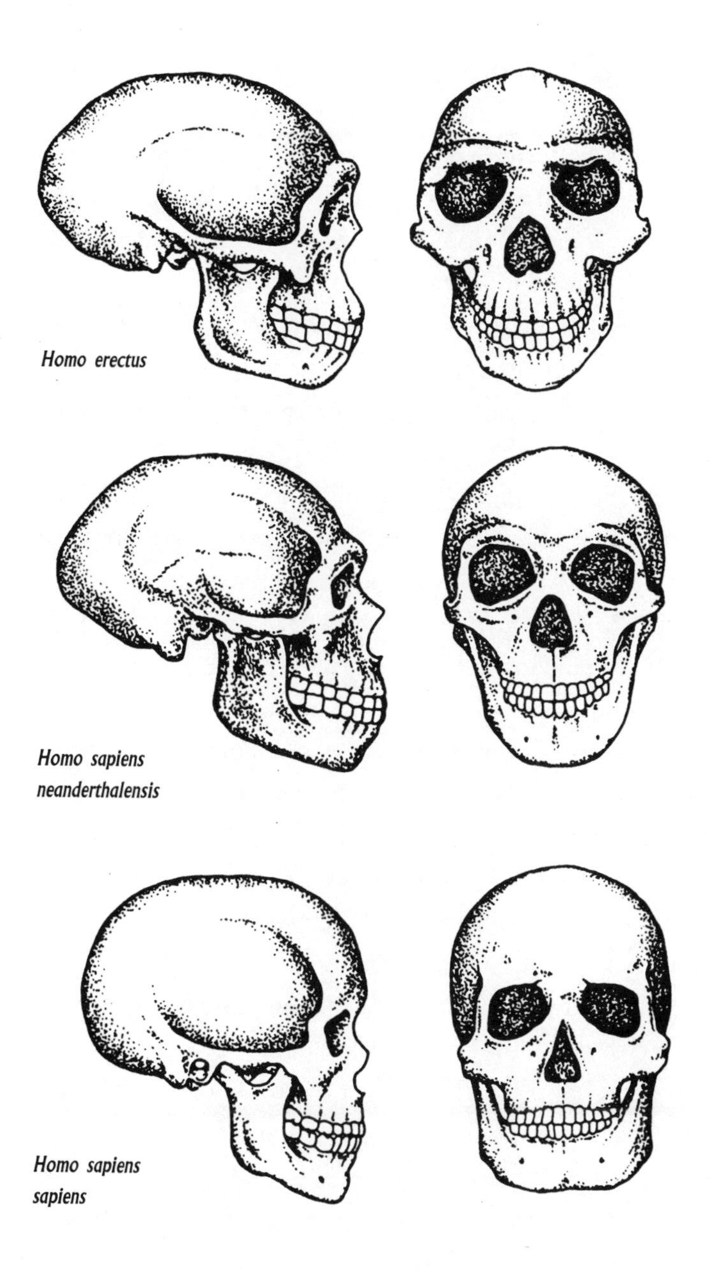

Moderns and ancient hominids. (Modified from Leakey and Lewin, Origins Reconsidered; with permission.)

African fossil beds deposited between 20 and about 15 million years ago, we have a rich record of these creatures, detailing a great evolutionary radiation of apes. And then the fossil record, at least for the traditional hominid hunting grounds of east Africa's Great Rift Valley, almost disappears, for there are few fossil hominoids known from this area in sedimentary rocks that date from 14 to 4 million years ago.

Africa was largely forested as late as about 15 million years ago, but then it too saw its great tropical forests shrink. Northern Africa gradually became drier, while lands to the east and south became regions of savanna and scattered trees. In this world the primates continued to evolve, and many modern monkey groups appeared in a widespread evolutionary radiation that occurred about 8 million years ago. In 4-million-year-old sediments, the first fully bipedal hominid fossils have been found. They represent a species formally known as *Australopithecus afarensis*.

From this point in time, the journey leading to *Homo sapiens* was relatively swift. The spectacular discovery of a nearly complete, 2.6 to 3.2-million-year-old hominid from Ethiopia by a team led by anthropologist Don Johanson (a fossil of a young female he affectionately named Lucy) filled in many of the gaps in our knowledge of human evolution. These oldest members of our tribe were far smaller than we are; the largest males weighed about 100 pounds. One surprising aspect was a striking sexual dimorphism: Males were between 50% and 100% larger than females. This feature suggests that, like many modern primates, the australopithecines traveled in troops rather than forming permanent family groups. The brain of these creatures was about 20% to 30% larger than that of a chimpanzee and about one-third the size of ours. It is apparent that a fully bipedal existence preceded what we would call a large brain.

There were several species of australopithecines existing between 2 and 3 million years ago in Africa. One of these probably gave rise to the first member of our genus, *Homo*. Primitive members of our own genus are distinguished from the australopithecines by a larger brain capacity, improved bipedal locomotor ability, and a shortened face. The oldest species of *Homo* has been found in African beds almost 2.5 million years old.

Africa is the well-known origin of humanity, and you can't enter a bookstore these days without seeing a display featuring a book by Leakey or Johanson or some anthropologist who disagrees with them: *Origins*, or *Original Lucy*, or *Origins Part 2*, or *Lucy's Original Origins*, all dealing with—surprise!—the origins of humanity and all based on finds from the Great Rift Valley of Eastern Africa. Most of the action revolved around the early parts of the family tree, back about 3 to 4 million years ago when we climbed up off our four legs, quit being *Australopithecus*, and started being *Homo*.

Less popular attention has focused on the final piece of the evolutionary puzzle: When did we quit being archaic *Homo sapiens* or even advanced *Homo erectus*, and become *humans*? How was that transition accomplished? Did we change slowly, over 100,000 years or more? Or was the transformation much more rapid? Did we become humans at many places, or are we all related to a single mother, dubbed Eve by molecular biologists, who lived some 200,000 years ago?

These oldest humans, for humans they must be called, gazed on a world perhaps not so different from the Africa of today. Surely they saw all the African mammals that are so familiar to us now. And surely one of the earliest human preoccupations had to be with the largest of all the African creatures, the elephants, mastodons, and mammoths: the three great lineages of the group known as the proboscideans.

. . . .

Humans and elephants go back a long way. Like us, elephants and their lineage are of African ancestry. Like our earliest ancestors, theirs survived the great Chicxulub impact of 65 million years ago. Those earliest members of the elephant lineage were the size of rats, as were our forebears; they had no trunks, great ears, or trumpeting cries. They began as small herbivores very much like the present-day cony. From the same group diverged the secoes some 60 million years ago. No creature even resembling an elephant arose until the Eocene Period, about 55 million years ago.

The oldest creature identifiable as an proboscidean is named *Moeritherium*. It was the size of a pig and may have led a semiaquatic life

SPECIES	**Wooly mammoth** *Mammuthus primigenius*	**American mastodon** *Mammut americanum*	**African elephant** *Loxodonta africana*	**Asian elephant** *Elephas maximus*
HEIGHT	9–11 ft (2.75–3.4 m)	8–10 ft (2.4–3 m)	10–11 ft (1–1.4 m)	8–10 ft (2.4–3 m)
WEIGHT	4–6 tons	4–5 tons	4–6 tons	3–5 tons
BACK SHAPE	sloping	straight	saddle-shaped	humped
FUR	dense	probably dense	very sparse	sparse
HEAD	high single dome	low single dome	low single dome	double dome
EAR	very small	unknown	large	medium
TUSKS	curved and twisted	sometimes two pairs	gently curved	gently curved
TRUNK TIP	1 short, 1 long “finger”	unknown	2 equal “fingers”	1 “finger”
TAIL	short	medium	long	long

Differences between mammoths, mastodons, and elephants. (Modified from Lister and Bahn, Mammoths; with permission.)

among bogs and marshes of ancient Africa, where its fossil record appears. This most ancient of true elephant ancestors seems to have had a trunk. It also exhibited another elephant characteristic, a reduction in the number of teeth, as well as the formation of tusks derived from incisors. Soon after the evolution of *Moeritherium*, the proboscidean stock diversified rapidly, and by about 30 million years ago mastodons had evolved. The first true elephants did not evolve until about 10 million years ago. Included among this group are our African and Asian elephants. Extinct mammoths, although technically part of this family, are here differentiated from true elephants.

People often confuse mastodons, mammoths, and elephants. Mastodons were the first of the three to evolve, though all three sometimes lived together during the Ice Age. Mastodons seem to have originated some 30 million years ago in what is now Egypt, and they differed from elephants and mammoths in usually having both lower and upper tusks. Their chewing teeth were also significantly different. Mastodons gave rise to the true elephants or at least shared a common ancestor with this group.

Two stocks of elephant are alive today: the genus *Loxodonta*, or African elephant, and the genus *Elephas*, the Indian elephant. *Mammuthus*, the mammoth, is closely related to the latter, which migrated out of Africa and diversified into many different species. All three appear to have diverged from a common ancestor perhaps 3 million years ago, and all had their earliest origins in Africa. *Elephas* and *Mammuthus* migrated out of Africa and eventually colonized most of the world during the last 3 million years. It is interesting that in one behavioral characteristic, the elephants resembled us: They began to wander literally to the ends of the earth. Why did they wander? Why did we? One thing is sure. The wanderings of humans and elephants reached epic proportions during the Ice Age.

. . . .

Starting about 2.5 milion years ago, the Pleistocene Epoch—or the Ice Age, as it is more popularly known—was a great trial by winter for our planet. More snow fell each winter than melted in the spring. Year by year this excess of snow and ice caused the formation of glaciers, which slowly crawled southward. Eventually, continental glaciers coalesced and merged with mountain glaciers, uniting to grip the land in ice and winter.

By no means was the entire planet covered by ice, as seems to be popularly imagined. There were still tropics, coral reefs, and warm, sunny climes that were pleasant the year around. But probably no place on earth, save the deepest sea bottoms, was completely unaffected; This is because the global climate changed, causing shifts in wind and rain patterns. Even places far from the ice became colder, some even grew warmer, and many became considerably more dry. Gigantic, cold deserts and semideserts expanded in front of the advancing ice sheets, while regions that had been dry, such as the Sahara of northern Africa, experienced increased rainfall. Conversely, the great rainforests covering the Amazon Basin and equatorial Africa, whose climates had been relatively stable for tens of millions of years, experienced such pronounced cooling and drying that large tracts of jungle retreated into pockets of forest surrounded by wide regions of savannas.

In North America, at least, the southward advance of the ice flows halted in the middle of the continent, and this maximum extent of the latest glaciation occurred about 18,000 years ago. To the north, most of the land was uninhabitable ice, whereas to the south, centuries of drought produced huge regions of desert and shifting sand dunes. It was an extraordinary time in the history of the earth. Even so, it was not unique, for glaciation affected significant portions of our planet many times in the remote past, such as during the Precambrian Era, some billion years ago, and most notably in the Permian period of 260 million years ago. Yet the last episode, which [ended] only 12,000 years ago, was one of the most intense.

The glaciers changed the nature of life on Earth and, in many regions, the geography of Earth itself. In Europe, the expanding and retreating ice sheets carved the fjords of Scandinavia as well as many of the features of northern Europe. In North America, the ice gouged Puget Sound in Washington State and the great inside passage stretching from southern British Columbia to Alaska. In mid-continent, the Great Lakes were created, and in Asia, huge lakes such as Lake Baikal were similarly wrought by moving ice. Lakes dammed by ice created monstrous spillways when the ice dams

Dinotherium

Mastodon

Mammoth

Ancient proboscideans. (From L. Figuier, La Terre Avant Le Déluge, 1864.)

eventually broke. Finally, the retreat of glacial sheets left huge piles of gravel and debris spread over large expanses of all northern continents.

It had long been postulated that the Ice Age comprised four separate glacial advances and retreats during the last 2.5 million years, the last ending only about 12,000 years ago. During recent years, however, improved chronology, derived from a combination of new methods for geological dating, has shown that the history of glaciation was far more complex and consisted of many individual glacial advances and retreats. At least 18 separate advances and retreats are now known. They have occurred at roughly 100,000-year intervals, and both the extent of the glaciers' encroachment and their size have increased through time.

As might be expected, the last of the glacial intervals left the newest and least disturbed geological record. This last advance and retreat, known as the Wisconsin glaciation in North America and as the Wurm in Europe, began about 35,000 years ago when glaciers started to increase in size both in mountainous areas and in the high latitudes of the northern and southern continents. It ended about 12,000 years ago, when these same glaciers finally melted away to approximately the positions they occupy today. At their maximum, the Wisconsin ice sheets covered most of Canada and extended far to the south in the American midwest. Great glaciers also grew out of the world's high mountains. The Pacific Northwest from approximately the position of Tacoma to Alaska was under ice. England, Scandinavia, Greenland, and much of the Baltic region of northern Europe were buried under a mile of ice.

The cause of these great glacial advances and retreats has long been debated. The primary agent is easy to pinpoint: The Earth became colder. The great ice caps began to spread from the polar regions, and as the ice caps spread, the amount of sunlight reaching the earth was increasingly reduced. But why was there a series of pronounced cycles, and why did the cooling begin and ice caps start to grow in the first place?

Several explanations have been put forth. Some scientists believe that the sun's energy output diminished, whereas others point out that the closing of the Isthmus of Panama, which occurred at about the same time, radically changed oceanic circulation patterns, bringing about a period of cooling. Paleontologist Steven Stanley has argued that this event brought on not only the Ice Age but, indirectly, the evolution of humans as well, because the change in climate that caused the ice ages is implicated in the environmental changes that may have prompted our evolution. Most scientists, however, suspect that glacial advances and retreats are caused by a far more gradual process: the changing distance between Earth and the sun.

Earth travels around the sun in an elliptical, rather than a circular, orbit. But the spinning Earth also resembles a giant gyroscope, and like that toy, it slowly wobbles as it spins. The change is almost imperceptible in the time frame of a human life. It is the tilt of Earth's axis that wobbles, and one revolution takes 22,000 years. The result is that the severity of summers and winters gradually changes, depending on the relationship between Earth's tilt and its distance from the sun. Summer in the northern hemisphere is likely to be hottest when the longest day of the year coincides with the Earth's being at the point in its orbit closest to the sun. Earth is in this position every 22,000 years. As if that were not enough, the tilt of our axis itself slightly varies over long periods of time, bobbing up and down between about 22 degrees and about 24.5 degrees over a cycle of approximately 41,000 years.

This changing relationship is called precession of the equinoxes. According to a theory first proposed by Yugoslav astronomer Milutin Milankovitch, the glacial advances were set off when the winters were coldest during the 22,000-year precession cycle, when the earth reached its maximum distance from the sun during the northern hemisphere winter. But there must be more to the story, for precession as been occurring for as long as Earth has orbited the sun, whereas ice ages have occurred infrequently in Earth history: 400 million years ago, about 275 million years ago, and starting 2.5 million years ago. The drift of the continents during the last 60 million years must have had much to do with the onset of the Ice Age as well. The southward drift of Antarctica to cover the South Pole was one factor, as was the arrival of North America and Greenland at their present positions by continental drift, for these movements effectively created a land-locked sea, the Arctic Ocean, covering the North Pole. Isolated from any warm ocean currents, the Arctic Ocean soon became covered by reflective pack ice and further cooled the planet. It may be that the separation of the continents and the creation of ice caps brought Earth to the threshold of glacial formation and that it was then pushed into the long Ice Age winter by precession-induced temperature regimes.

The convulsing ice and changing climate would at first seem to be agents of decreasing faunal diversity, for huge areas of formerly productive land were entombed by the steadily creeping ice. But some forms thrived. Many land areas bloomed under the new conditions, and newly evolved vegetational regimes often favored large, herbivorous grazers. For example, great forests often seem like oases of life but in fact are poor places for sustaining populations of larger herbivores. Many trees have evolved complex defense systems that discourage browsing of their limbs, and the overall density of many forests reduces the mobility and size of larger animals. It is the more open environments, between forest and grassland, that are usually the most favorable habitats for larger herbivores. With the onset of the great glaciers, many previously forested regions, such as the rain forests of the Amazon Basin and large tracts of North America and Africa, grew more patchy in distribution or became savannas of mixed grassland and shrubs forming a complex vegetational mosaic. In response, new, larger herbivores evolved. Thus was born the last great assemblage of larger mammals, the Ice Age megafauna.

The time of the Ice Age is of utmost importance to humanity, for it is the time of our origin. We began this interval as australopithicines, ape-like forms living and dying among the other wildlife of Africa. We ended the Ice Age, only 10,000 years ago, as humans, living on every continent except Antarctica. For humanity, the Ice Age was the crucible of evolution.

. . . .

The sudden onslaught of glacial ice some 2.5 million years ago, and the creation of markedly different climates that accompanied this new Ice Age, completely changed the selective pressures acting on thousands of species living on Earth. These pressures fostered species adaptation to the new conditions by favoring the reproductive success of individuals that inclined toward traits such as hair growth, larger size, and the gradual production of physiological features that enhanced survival in a radically changed world. Evolution was, as always, the ticket to survival, and at the pivotal point of the Ice Age, new and divergent paths seem to have opened everywhere.

THE EMERGENCE OF HUMANITY. (FROM L. FIGUIER, *LA TERRE AVANT LE DÉLUGE*, 1864.)

5

Wheel of Fortune

THE GREATEST INTELLECTS PRODUCED BY our species have succeeded in piecing together a sketchy picture of how the simplest particles of matter interact. We now have a vague idea of what light is, how hydrogen atoms attract and repel one another, and how matter behaves. It has taken untold wealth and many centuries to achieve this view of the physical world at its least complex level. The study of matter and energy and their interaction is called physics. It is a testament to the intelligence of physicists that they are smart enough to study the simplest systems in our universe.

Life, on the other hand, is certainly the most *complex* assemblage of matter in the universe. The study of how varieties of living organisms change over time can be called evolution. This study has been going on only for about a century and a half. Yet evolutionists are castigated because they have not made as much progress with their DNA molecules as physicists have with their hydrogen atoms.

When systems get really complicated, it is far easier to invoke the supernatural than to do the hard work of scientific explanation. Evolution is so complex that it is easier to ascribe it to God, but even so, why should fear be involved?

Why is the theory of evolution so feared by evangelical Christians in the United States? Is gravity to be feared? Does magnetism inspire assaults by angry parents on school boards? Why does evolution, no more or less a theory than these other accepted tenets, seem especially godless and menacing to otherwise rational people?

Is the state of Washington, where I work, I was recently told by a professional teachers' group that only half of high school science teachers in rural districts in my state even believe in evolution, let alone teach it. And why is it that only Christianity so fears an ancestry in as noble a lineage as the primates?

Maybe we can throw some light onto this question by turning evolution into a game. We need a grasp of evolution, after all, if we are to understand the deadly duet of evolution and extinction played out by humans and elephants.

. . . .

Let's examine evolution by turning it into a board game. For want of some better term, we might call it the Game of Evolution. Personally, I prefer the name Wheel of Fortune or perhaps Casino (both unfortunately taken already), because evolution is so much governed by chance.

As in all classic board games, we have to move from some starting point to some end point, and in the process we have to undergo change—or, in this case, to evolve. In Monopoly, for instance, we go around and around the board, and as we complete the circuit, our financial condition is altered: We grow richer or poorer as hotels and property are developed.

Our board game may have a lot in common with Monopoly. For instance, Monopoly takes a long time to play. I can't remember the last time I finished a game, although I remember starting many. *Our* game will even take longer—on the order of hundreds of millions of years.

What are the rules? Here things get trickier. There are any number of ways in which we could design our game, because the real game of evolution is played in many ways, just as there are many ways to interpret who wins and who loses. So we have to design several versions of the game.

In our first version, we can equate *surviving* with winning—if you last 500 million years without your lineage terminating completely, you win. But there will be many ties in this game. Perhaps a better way of deciding on a true winner among survivors is to base winning on diversity: The player who has produced (or evolved) the most surviving species wins. I therefore propose that every player start out as some simple creature that must then go through time, trying to avoid extinction and at the same time producing new species. Herein lies the fun. To get around the board, which is equivalent to passing through eons of geological time, you generally must adapt or die—because, as in any good board game, conditions change from square to square. To adapt, you must evolve. And evolution, of course, creates new species.

Let's say that each of us starts out as some simple sea creature 500 million years ago, and the end of the game is reached when (if!) we and our evolutionary descendants survive to the modern day.

Players can employ numerous strategies to increase their chances of survival. One approach is to live in an environment that never changes. In such circumstances you don't need to evolve. You can stay as you are, through thick and thin, and hope that you already have evolved such a superior body plan that you can weather all the changes that Mother Earth is going to go through in the 500 million years of the game. Never mind that no known species have existed for 500 million years, although there are some slightly higher categories, such as genera, that may have lasted that long. We call these few survivors living fossils, and we celebrate their antiquity. About a dozen or so come to mind, such as the brachiopod *Lingula*, cockroaches perhaps, the Nautilus, and a few others. But these are very rare examples. The species known from the fossil record rarely lasted 5 million years, so it is a good bet that the strategy of simply not evolving may not be the best choice.

Perhaps evolving is a better way to go. Why evolve? Because conditions on Earth have a nasty habit of changing. Continents shift, climates grow hotter or colder, oceans turn more or less saline, and the gas composition of the atmosphere changes. It would be great to start our game with a physiology that is tolerant of all such changes, able to live on land or in the sea, to thrive at high or low temperature, and to eat either meat or vegetable matter. Any organism so broadly adaptable is unlikely to be so fit for any *one* environment as to thrive in competition with more narrowly adapted species. Unfortunately, no creature on Earth today has such a range of adaptation, and certainly nothing that was around in the Cambrian period, when our game starts, was so lucky. So that option is very risky, with low odds for survival. Instead, either you must be more narrowly adapted (a "specialist") and try to live in environments that don't change, as we have noted, or you must make yourself numerous and geographically widespread. In this way you might squeak through, somewhere, when old Mother Earth throws some curve ball at the biota, such as a sudden onset of glaciation or a long-term change in sea level.

Widespread distribution does seem to be a preferred route to survival. Why, then, can't every player simply opt for high numbers and worldwide distribution? Sorry. In our game, as in real life, the amount of resources (in Monopoly it is the money; in our game it can be food or living space) available is finite and must be shared—or fought for. Let's say you can be a specialist or a generalist; specialists rake in much more food every time they pass Go (so not everybody collects $200), whereas generalists receive less money but get to be more widely distributed. Each of these "choices" exacts its own consequences. Certain conditions (such as rapid change) favor generalists, others favor specialists, and you never know in advance which choice will be better.

Now, to play the game you have to go around the board (which in our game will be equivalent to going up through time), and that involves rolling the dice. This is very appropriate. In our game, even the best-adapted plans (body plans, in this case) can fail (suffer extinction) at the whim of the dice. In this, our game mimics nature. For instance, let's assume you have everything going for you, but then a great meteor hits Earth and you are incinerated. Survival is a risky business, and evolution doesn't provide for elaborate advance planning. You can put on a winter coat of fur and store up body fat for a coming winter, but only humans can really do long-range planning. And even in humans, this ability seems to be limited, judging by the state of the world today.

Let us begin. I know, we have not fleshed out all of the rules, but who ever reads all the instructions anyway? Let's play, instead, and go for on-the-job training.

We begin at the end of the Cambrian Period, some 500 million years ago. Every player starts out as one species, and the player whose lineage is represented by the most species alive on Earth at the end of the game (the modern day) wins.

I have chosen to be a cephalopod mollusk. With a roll of the dice I march three squares. One, two, three . . . lands me on the Lower Ordovician, some 450 million years ago. Landing there, I get to pick up a card from the pile, telling about conditions at that time. Let's see: Widespread flooding of continental interiors produces large shallow oceans (good for me), formation of coral reefs (neutral for me), widespread diversification of crustaceans and trilobites. Thus if you are a creature with swimming ability (I am), jaws (that's me) and thrive in normal marine salinity (yes!), you get to diversify into a thousand new species (Eureka! I love this game!). In one fell swoop I am the most diverse carnivore in the oceans and the richest (in terms of diversity) player in the game. I have to laugh at the player on the square behind me, the trilobites. The late Cambrian extinctions have wiped out almost all of them, and with each roll their numbers dwindle further. But it is a heyday for us good old nautiloids.

My turn again. Roll 'em . . . box cars. Whoops, that lands me directly on the Upper Devonian period—a large red square—global mass extinction! An ice meteor hits planet, wiping out 60% of all species, and worst of all, *any* species's fate in this mass extinction is completely related to chance, not fitness. So I take a card, look up nautiloid survival rates in the Upper Devonian period mass extinction, and find that my fate is . . . not too bad compared with everybody else. I lose about half of my species, less than the global extinction rate. There are far worse losers. The brachiopods almost go extinct.

But my next roll is a disaster. I land on a 350-million-year-old square that says fish have evolved, and that they not only compete for my food resources but do a better job of it. In a resource-starved world, it really is survival of the fittest. I lose another 40% of my species, and I still have to get through the 250-million-year old Permian mass extinction, the end-Triassic extinction, and the great Cretaceous–Tertiary mass extinction. Each of these wipes out at least 50% of all species on earth, and the end-Permian disaster takes out as many as 90%. In each of these catastrophes, the great stock of species I had accumulated early in the game dwindles, and although I continue to produce new species, I do so at a diminishing rate, because I have a design built for the good old days of 500 million years ago. In the increasingly modern world, I am obsolete and getting more so. My extinction rate now far exceeds my species origination rate. By the end of the game, in the modern age, I am down to four species: the modern nautiloids found only in the South Seas. The player next to me—insects—now proudly boasts 30 million species! I lose. But there are far worse losers. Where are the trilobites, ammonites, graptolites, and thousands of other denizens of the Paleozoic and Mesozoic? Extinct. These are the big losers. At least I am alive at the end of the game, if in greatly reduced circumstances.

To be honest, this version of the Game of Evolution isn't very accurate, nor does it tell us anything about life, other than that life had a history and that much of that history was shaped by chance. Perhaps winning should be based not on species diversity at the end of the game but, like Monopoly, on how rich you are in resources. In Monopoly you win by getting all the money on the table. In the game of evolution, you win by reaping the lion's share of our planet's food and space resources.

By this new criterion, the hominoid primates (a real loser at the end of the last game—they finished with only a *single species*) would be the winners, hands down. Although some might argue that the big winners in today's world are the insects, which vastly outnumber all other species combined, can it be said that they appropriate the planet's resources to the extent that humans do? Hardly. We humans invented DDT, not the bugs. Resource domination goes to *Homo sapiens* in this game. And to this winner, in my view, goes an even greater prize: species immortality.

It is my opinion that no matter where on the board we humans land and no matter what card we draw, we cannot be knocked into extinction. Epidemic, meteor hit, Ice Age, global warming, or any serious plague that could knock most species out of the game can no longer touch us. With a population approaching 6 billion—and even more important with a distribution that reaches from the equator to the poles and from the flatlands to the tops of the highest mountains—we are by far the most widely distributed terrestrial species on the planet and, with our technology, I believe the most unassailable.

What could drive us extinct? There is a current flap about microbial plagues, yet such plagues rarely drive any species into extinction. There is a real (but fading) threat of nuclear Armageddon, yet it would require a truly global nuclear exchange to wipe us out. Perhaps global pollution will kill off our crops, yet such pollution would have to be toxic beyond anything known on Earth today to kill off *all* plants, and we can eat almost any living thing and survive. Perhaps the most dangerous threat is that of an enormous meteor or comet striking Earth or a solar accident of unforeseen intensity—and who can say that in the next half-millennium our species will not make itself immune even to those remote threats through interstellar space flight. We are the least endangered species on the planet.

We humans have apparently won the real game of evolution. The game is over until our species disappears. That, too, seems to be one of the rules of the game. Once there is a winner, the game is over. There is no playing for second or third place.

. . . .

Let us try another variation and look more closely at the role of chance in the Game of Evolution. This time, instead of using a board game, let's pretend that the history of life has been recorded on some long-playing tape. What if we could replay that history: rewind the tape, do it over, and see if things would occur again exactly as they did? Would we have another Age of Dinosaurs, another *Archaeopteryx*, another coelacanth, another human?

In this simple thought experiment, we make every single physical event during the thousands of millions of years of Earth's history repeat itself. For instance, we exactly duplicate the motions of past continents as they slowly moved over Earth's surface on their great lithospheric plates. We make the times when the world was covered with glaciers exactly the same. We make increases and decreases in sea level, temperature changes, and atmospheric proportions of oxygen and carbon dioxide identical. We create that most favored of science fiction scenarios: a parallel universe. Then we unleash DNA and stand back. Would evolution proceed in exactly the same way it did over the last 3.5 billion years of the history of life on Earth? Would we have the same assemblage of creatures we have now? Would monkeys still look like monkeys and would all of the fish and beetles and so much else be identical? Or would some differences be discernible? Perhaps we would have something strikingly different. Would we have zebras without stripes, or eight-legged ants, or a world without flowers this time around? Would the Age of Dinosaurs have happened? My hunch is that even if we made physical conditions exactly the same in each run of our long-playing tape, the biological results—shaped by the forces of evolution—would yield a different assemblage of animals and plants every time. Perhaps they would not be *very* different, but there is a great deal of chance in evolution, such as that involved with mutation and with survivability in the small populations from which all new species arise. The Nobel laureate Murray Gell-Mann, in his recent book *The Quark and the Jaguar*, describes evolution as being the result of basic laws of physics, plus randomness, operating under natural selection. At every moment there is a different road to take.

It is a common human trait to take the familiar for granted. Consider the wondrous diversity of mammals now living on our planet. The forests, the grasslands, the oceans, and even the sky have been colonized—and in a sense conquered—by mammalian species. There are so many mammals, and they are so dominant, that geologists have long called the present geological era the Age of Mammals. And yet this "Age of Mammals" has been the order of things for but a small fraction of life's existence on Earth. The fossil record tells us that mammals have existed for 250 million years but have been dominant for only the last 50 million years. Perhaps we should ask why the Age of Mammals has lasted such a *short* time or why mammals now rule the earth at all.

We humans (good mammals ourselves) have been indoctrinated with the chauvinistic notion that mammals are the dominant land fauna because we are *superior* to the other classes of land vertebrates: the birds, reptiles, and amphibians. Birds are clearly the dominant creatures in the air, but only a few exist as purely land-living forms; and compared with mammals, reptiles and amphibians are relatively scarce in most terrestrial habitats. Most scientists agree that the aspects of "mammalness" that give us our competitive superiority are a warm-blooded metabolism, a highly developed brain, a good fur overcoat to withstand cold winters, and parental care of the young.

When I was in school, the story of the rise of mammals went something like this: The first land-living creatures with backbones crawled from the sea about 400 million years ago. These earliest terrestrial pioneers were amphibians. They were the evolutionary ancestors of reptiles (distinguished by the ability to lay eggs and grow from juveniles to adults in a fully terrestrial habitat, as well as by differences in skeletal architecture from the original amphibian designs). The reptiles in turn gave rise to the first true mammals somewhere about 225 million years ago. The early mammals did not immediately take over the earth but had to serve a rather lengthy apprenticeship to the then-incumbent terrestrial overlords, the dinosaurs. Eventually, however, quality prevailed, and the dinosaurs became obsolete. Succumbing to pressures of climate change and the onslaught of hordes of little egg-eating mammals, they went extinct against a backdrop of exploding volcanoes. In this scenario, the dinosaurs had the decency to hand over the stewardship of the continents to creatures better adapted to living on land. In short, mammals now control the earth because they are inherently superior. Or so the story goes.

Let us re-examine this scenario by conducting a paleontological thought experiment. We are going to reintroduce dinosaurs to California.

California once had many dinosaurs. Their fossils are not common, but now and again sparse dinosaur bones occur in the Cretaceous strata of California—enough to tell us that dinosaurs did live in the Golden State long ago. With our time machine, we can revisit that place and muse about what that world may have looked like.

The beauty of thought experiments is that you don't have to worry about your budget, so let us spare no expense. Let us build a giant fence around the entire northern half of California and introduce a 70-million-year-old (Late Cretaceous) dinosaur fauna into our habitat. We now have big dinosaurs, such as *Triceratops*, many duck-billed dinosaurs, a few long-necked sauropods, lots of smaller dinosaurs, and of course a few large carnivorous forms such as *T. rex*. Northern California has a diversity of habitats, ranging from the great redwood forests in the northwest (which would certainly be familiar to our dinosaurs) to the dry grassland of the Sacramento Valley and the Sierras in the east. Surely there would be plenty of familiar food and habitats for our herbivorous dinosaurs (although grass would be a novelty and might pose a challenge). Thus it is quite possible that our new immigrants would survive at least for a while; perhaps they would even thrive. But our park is not solely populated by dinosaurs, for we are also leaving within it the native mammalian populations that exist there today, such as deer, elk, porcupines, bobcats, bears, and a mountain lion or two. The only indigenous Californians *not* invited are we humans; all of our technology and works have been omitted as well. Thus our great northern California dinosaur park would be stripped of all humans and of the shopping centers, cities, and especially freeways that make California such a pleasure today.

This scenario conjures up fantastic visions. Imagine the first encounter between an angry momma bear and a hungry *T. rex*, or picture a mountain lion coming across a small *Triceratops*. Would the dinosaur predators find the mammals not only succulent but also easier to kill than their normal ceratopsian prey? Would rodents find dinosaur eggs just too good to pass up? Interesting images indeed.

But our experiment is not about only the short term, or the more spectacular scenes and struggles; we are interested in a longer view. Let us leave our great park undisturbed for 10,000 years to allow this mix from the Age of Reptiles and the Age of Mammals to do its thing. In evolution, time is the great leveler, the true arbiter of adaptation and fitness. Evolution is an endurance race, not a sprint.

What species would we find when we came back after 10,000 years? Would the mammals, with their parental care and warm-bloodedness, have won out, having driven the dinosaurs to extinction through competition for space and food? Or would the dinosaurs, perhaps warm-blooded and good parents as well, have held their own? Would some curious but wonderful combination of mammals and dinosaurs greet our return? My guess is that the dinosaurs would successfully compete with mammals for food and space and that they might just wipe out most of the larger mammals over time. Perhaps our California dinosaurs would even escape from their California prison and (like good Californians of today) move to more "livable" places, such as Oregon or Montana, eventually coming to rule the earth once more. Michael Crichton may be right: At any cost, keep the damn things off the mainland.

Now let us repeat this trick 50 times. Would the assemblage of animals still alive at the end of the experiment be the same each time? Would the experiment repeat itself exactly at each iteration? I think it most unlikely. I suspect that, like human history, the history of life is susceptible to minor shifts that are amplified into gigantic changes over the passage of time.

If they were so well adapted, why are dinosaurs no longer around? That we know: the Chicxulub comet. The dinosaurs had been carrying on quite nicely for over 100 million years prior to this comet strike. Had that great impact not taken place, we would surely not have the animals and plants now on Earth today, and there is a very good chance that the dominant land animals would still be dinosaurs. And if that had been the case, would our species have evolved? I think not.

Let us set up our experiment another way. Putting dinosaurs in California is fun, but the stretch of time between the last dinosaur and today is enormous, about 65 million years. When dinosaurs last walked the earth, ecosystems were vastly different from those of today; for instance, at the time of the last dinosaurs, such common plants as the grasses had not yet evolved. Maybe dinosaurs wouldn't do well in our world after all, but for reasons completely unforeseen and unrecognized—reasons as minor as the pathways of energy through the terrestrial systems or the presence or absence of digestive bacteria in their guts—things invisible to the fossil record. So let us rerun our experiment with a somewhat more modern assemblage of extinct creatures. Let us replace all the dinosaurs in our great California park with Ice Age megamammals, the same assemblage that lived in California as recently as 12,000 years ago. Let us add mammoths, mastodons, saber tooths, camels, horses, and all the rest. Again, we will allow no humans to enter our great cage as they did some 12,000 years ago, when this experiment actually played out. Let the whole thing percolate for 12,000 years and open up the box. What animals are still alive? I would put my money on the mammoths.

. . . .

Evolution isn't rocket science. Every new type of creature originates through the process Darwin called speciation. In most or all cases, that origination takes place in one geographical area and among one, usually isolated population. Then, if the new species has attributes that enable it to compete successfully within its ecosystem, it thrives. And spreads.

It was long fashionable to think of evolution as acting as a progressive, directional force. Sometimes it is—such as when lineages of animals evolve increasingly larger sizes over time. But our newer views of the fossil record suggest that evolution is not continuous but rather occurs in fits and starts, and it is certainly not progressive. Much of what we humans see as "directed" evolution may be colored by our own values and our reluctance to embrace a stochastic and random world. For example, how did a wondrous structure such as an eye evolve, if the process were not somehow directed by natural selection or even a supreme being? And doesn't evolution tend consistently toward more complex animals and plants? No. Organisms get smaller as well as larger, simpler as well as more complex. We do not see the failures as often as the successes, because they do not last so long, and it is hard to acknowledge that there is a great deal of chance even in such classic examples of progressive or directed evolution as the tendency toward ever-greater brain size in hominids. Of all evolutionary trends, increase in body size is by far the most common. Paradoxically, it tells us almost nothing about how evolutionary change and the creation of new species actually occur.

The great breakthrough of Niles Eldredge and Stephen Jay Gould—their 1972 theory of punctuated equilibrium—proposes that most morphological change occurs during the speciation events, when a species originates, and that virtually no further evolutionary change occurs thereafter. The forces that produce new species are brought to bear when small populations of an already-existing species get cut off from the larger population. Once the two populations can no longer interbreed, gene flow—the interchange of genetic material that maintains the integrity and identity of any species—is cut off. The elimination from the gene pool of the smaller body of individuals has no effect on the larger population, but the smaller population is almost always drastically affected when the larger gene pool is no longer available to it.

Imagine, for instance, that we take a population of a million humans with equal numbers of every known race, put them on one land mass, and allow them to interbreed freely. Eventually, all traits that are characteristic of any single race will largely disappear. Before that point, however, let us remove ten members, randomly chosen, from the larger population. The removal of these ten individuals has no effect on the gene pool of the larger population or on that of its offspring. However, being cut off from the large population tremendously affects the overall gene pool of the poor isolated ten. Furthermore, the ultimate genetic makeup of the offspring of these ten will itself be enormously influenced by the content of our founding ten. If all our now-stranded humans happen to be Caucasians, or if all are mongoloids, or all short or all red-haired or all whatever, we will very quickly get a gene pool made of up gene frequencies that are very different from those of the larger, parent gene pool. The influence of the chance assortment of genes in a small, isolated population is known as the founder effect, and it seems to play an important role in species formation.

Add the last part of the mix—new environmental conditions different from those experienced before—and you have a recipe for making a new species, given enough time and given continued separation of the two groups.

The keys to all this are that (1) the two populations remain physically (or behaviorally) separated so that no interbreeding takes place, and (2) each has to experience and deal with different environmental conditions or challenges that affect the survival and reproductive viability of offspring. With these caveats in place, it is very unlikely (though it is still possible) that our separate populations will evolve in exactly the same way. Over time, two separate species may emerge, such that no amount of interbreeding will produce offspring.

The world now has 1.6 million identified species and surely millions more not yet discovered. Some experts on biodiversity, such as E. O. Wilson of Harvard University, estimate that there may be more than 30 million species on Earth. Add all the species that lived in the past and you have an immense number. The production of new species is apparently not difficult. What *is* difficult is to form a new species without some sort of gene pool separation. It is also hard for any type of evolutionary change to occur throughout an entire numerous and widespread population.

New species have formed many times in the course of human evolution. Although there are numerous gaps in the record, many disagreements among specialists, and much work left to do, we know the rough outlines of human evolution. The human family, called the Hominidae, seems to have begun about 4 million years ago with the appearance of *Australopithecus afarensis*. Since this origin, our family has produced as many as nine species, although there is debate about this number. About 3 million years ago two new species arose, *A. africanus* and *A. aethiopithecus*, and *A. boisei* appeared about 2.5 million years ago. (These latter three species are also sometimes identified as *Paranthropus* instead of *Australopithecus*.) But the most important descendant is the first member of our genus, *Homo*, a species called *Homo habilis*, which is about 2.5 million years old. This creature gave rise to *Homo erectus* about 1.5 million years ago, and *H. erectus* gave rise to our species, *Homo sapiens*, either directly, about 200,000 to 100,000 years ago, or through an evolutionary intermediate known as *Homo heidelburgensis*. Each of these changes involved a species formation event. Our species has been subdivided into a number of separate varieties, one of which is the so-called Neanderthal. Some workers even consider the Neanderthals a separate species, *Homo neanderthalis*, but this is highly disputed.

In 25 years of searching the fossil record, I have seen overwhelming evidence that the theory of punctuated equilibrium as originally described by Niles Eldredge and Stephen Jay Gould accurately describes the way evolution normally proceeds. This leads to several inescapable conclusions about the course of *human* evolution, as well as to predictions about the possibility of future evolution within our species. First, the formation of any new human species occurred when a small group of hominids somehow became separated for many generations from a larger population. Second, following the rapid morphological transformations that took place during the speciation process (let's say that this stage lasted a hundred generations), *Homo sapiens* showed *little or no further evolutionary change*. Oh, we changed skin color repeatedly, and stature, and facial morphology, and there was much evolution in response to infectious diseases, but all humans today can interbreed regardless of these minor styling details. The engine and body work are the same, as is the central microprocessor in all of us. We are different-colored (and different-shaped) peas in the same pod. For all intents and purposes, our species is through with major evolutionary change.

Rats! No giant brains for us! As a boy I once watched a great *Outer Limits* episode in which some machine accelerated the evolution of whatever luckless creature was thrust into it. When our human hero inevitably gets put into this evolutionary oven and baked, he comes out—no surprise—with a monstrous head and a *much* smarter brain. It is an image familiar to us all: Our future evolutionary development must entail further enlargement of our brain, further development of our intellect. It is soothing to think that evolution will boost our intelligence further or even that some "directional" force is guiding that development. Yet unless the "smartest" people have the most babies, the future intellectual development of humanity via a wholesale increase in brain size is unlikely.

What would it take to create a species of humans with, let's say, a brain size of about 2000 cubic centimeters, compared with the average value of 1100 to 1500 cc in *Homo sapiens*? This is not a trivial point. Human babies now undergo so much brain development outside the womb that they are more helpless at birth than any other mammal, and take longer to become self-sufficient. (What other animal takes a year to walk?) Yet the birthing process is also the most difficult. Brain size could not increase without either a concomitant increase in birth canal size or a much longer period of helplessness after birth, to let all the connections form in the larger brain.

All of this musing suggests a sobering fact: *Our* species's days as major evolvers are pretty much over. I suspect we will all be dark-pigmented within several millennia, because we are stripping away the ozone layer fast enough to kill off, with skin cancers, most Caucasians in the next few centuries. But apart from such minor details, our die is pretty much cast—and has been since the formation of our species. That is another tenet of the Eldredge–Gould concept: Since our formation as a new species some 125,000 to 200,000 years ago, we have shown little structural evolution. People were probably about as natively smart then as now. The various human races have slightly differentiated as our species has dispersed around the globe, but our brains have not gotten bigger since the Ice Age; we have not evolved more intelligence. That is why I believe that a human snatched from southern Africa or Ethiopia or any other locale of 115,000 years ago, immediately after our species first evolved, would surely look—and probably think—like us.

The hope that we have been continuously evolving is the rationale behind one of the great failed hypotheses about human evolution, the candelabra theory. It was long thought that *Homo sapiens* independently evolved from widely separated populations of *Homo erectus* at many places around the world, such that each "candle" on a many-branched candelabra would eventually be lit with the flame of our species. This was not the case. Both fossils and DNA evidence now show that only a single branch—a small population—evolved into our species, just as Eldredge and Gould predicted. This new species then took over the world.

A case in point comes from the fossil record of Europe. It was long thought that *Homo sapiens* evolved from coarser, shorter, heavy-jawed progenitors: the Neanderthals. These lumbering "cave men" have been found in a variety of locales in Europe and the Middle East, and they seem to date to between 200,000 and 300,000 years ago, although in a strict sense European Neanderthal populations are thought to have evolved between 200,000 and 100,000 years ago. According to the candelabra theory, the Neanderthals were but one branch of archaic humans that slowly evolved into true humans not only in Europe but also in disparate areas of Asia and Africa.

Recent discoveries from the fossil record and from DNA analysis have thoroughly debunked the candelabra theory. Species do not evolve simultaneously all over the earth. They do so in one locality: one isolated glade or water mass or mountain valley. True humans and Neanderthals may both have evolved from a common ancestor, a poorly known creature called *Homo heidelburgensis*, which itself was an African or European descendant of the far more archaic *Homo erectus*. Yet each of these events took place in but a single locality. The place of origin of these emerging species is not known, but it may have been some isolated region in the Middle East or, more probably, in that womb of humanity, Africa.

. . . .

Imagine the first population of air-breathing amphibians, and consider how this particular adaptation, so likely to enhance success in virtually any terrestrial environment, led to the proliferation of this particular species. Now imagine the success of a *people* with the most efficient and inquisitive brain ever evolved. *Homo sapiens*, or the Moderns, we might call them—a new species with a brain that would eventually design and build computers, conceive of black holes, create *Wheel of Fortune*. This too is an adaptation that must have been wildly successful. And these people, like the first amphibians, began to spread outward and away from their center of origin.

Where did this first population arise? The Moderns are derived from a more archaic variety of our species (often referred to, cleverly enough, as the Archaics, although they are more properly called *Homo heidelburgensis*), which appears to have originated about 200,000 years ago. Somehow, somewhere, a small band of Archaics became separated from the larger population. Was it a band in east Africa, or a group finding its way to north of the Sahara regions, alternately savanna and desert during the climatic oscillations of the Ice Age? Was it an isolated band in southern Africa or even Europe? In whichever locality, isolated for whatever reason, this band of creatures was exposed to some complex of environmental conditions that transformed it, over countless generations, into more gracile humans with less prominent brow ridges and larger brains—us.

Our start may be tied to a great climate change that occurred about the time we first evolved. The Moderns seem to date back to about 130,000 years ago, a time of one of the most profound climate changes in the history of this planet. The northern hemisphere was locked in ice 130,000 years ago; this interval of time corresponds to the Riss glaciation event of Europe and the Illinoian glaciation event of North America. By 125,000 years ago, however, the world had far warmer climates, and the transition from cold to warm, accompanied by rising sea levels and fluctuating climate over a few thousand years, surely perturbed the ecosystems of Earth in ways we can barely comprehend. From 130,000 years ago to 80,000 years ago the climate became warm, and it then remained warm and moist for a long time. During this period the Moderns escaped from their ancestral homeland and began to spread to the four corners of the earth. Our species ended up on every continent except Antarctica long before it could either forge metal or sow a crop. Long before the first cave painting, we had mastered boat building and sailed oceans to colonize Australia.

Why did we wander? Why leave Africa or even the Great Rift Valley? Was it a willing exodus, or were we actually expelled? Perhaps, we wandered for the same reasons many other animals do; we set out in search of a better life, for less contested food or shelter, for more safety from the great cats and jackals and canines of our ancestral home. Or maybe we wandered simply out of curiosity about what lay over the next horizon. For whatever reason, wandering was part of our heritage. If not provoked by them, perhaps our wanderings were aided by the climatic fluctuations of the Ice Age.

. . . .

In geology we revere the principle of uniformitarianism. This generalization, first proposed by 19th-century British geologists, states that the present is a key to the past. When applied to our efforts to understand the Ice Age, however, uniformitarian thinking may be leading us into errors of interpretation. The 2-million-year period, known as the Pleistocene Epoch, or Ice Age, is very unlike the rest of the vast stretch of time since the demise of the dinosaurs. For 63 million years following the great meteoric impact in the Yucatan Peninsula, climate was a predictable thing. For the last 2 million years, climate has been anything but predictable. There have been as many as 27 separate cycles of rising and falling global temperature in the last 2 million years. During the periods of cooler temperature, ice has crept from the north and out of the mountains and has tied up significant amounts of the water found on our globe. When the ice sheets have grown, the seas have dropped. When temperatures have warmed and the ice has melted, the seas have expanded and their level has risen. During the coolest intervals, the level of the sea was as much as 400 feet lower than it is now.

A 400-foot drop in the level of the ocean creates extraordinary circumstances. Even if it took a thousand years or more to drop such a distance, the receding shoreline would be noticeable in the lifetime of a given family, and the lore of "the good old days," when such and such was an island though it is now entirely in contact with land, would surely be passed on from generation to generation. Imagine our world if the sea were suddenly to drop 400 feet. Long Island would no longer be an island but part of New York and Connecticut; the barrier islands of the Gulf Coast and the Carolinas would no longer be islands. Far more striking would be the enormous enlargement of the coastline itself, for it would expand outward by tens of miles in most places.

Yet even these changes would be minor compared with those in many other parts of the world. Alaska would not be separated from Asia but would be connected by the famous land bridge, Beringia. England and Ireland would be part of a vast peninsula extending out from France. Africa would be more accessible from Europe, for vast regions of the Mediterranean would have disappeared.

To the east, even greater geographical changes would be apparent. India and Sri Lanka would be linked; the great continental shelf off China would be completely emergent; Japan would be connected to the mainland of Asia; and Australia and New Guinea would be one large continental block, separated from the Asian mainland by narrow straits via the Sunda shelf of what is now Sumatra, Java, and Borneo.

During the radiation of the human species, these conditions would have provided low-lying areas rich in water and plant life and surely rich in game. The expanded coastlines would have been ideal corridors for the long treks that allowed our species to conquer the globe.

The sea was at a minimum level about 130,000 years ago and again about 18,000 years ago. Other, lesser rises and falls also occurred, with notable low stands occurring at about 108,000 and 88,000 years before the present. Each of these intervals of low sea level made possible human migrations that could not occur today. The entire period from about 75,000 to 18,000 years ago saw sea levels at least 150 feet below current levels, and sometimes much lower.

The cycles of alternating warm and cold climate and the resulting rise and fall of the sea were surely important for the movement of our species. Modern humans, originating in Africa perhaps 130,000 years ago, reached Asia Minor 100,000 years ago and eastern Asia 70,000 years ago, and from there the first great sea voyages took our species to Australia about 40,000 to 50,000 years ago. Europe was not reached until about 40,000 years ago, and northern Asia only 25,000 years ago. The last great step to the "New World"—North and South America—did not occur until about 12,000 years ago.

When we arrived in a new place, what happened? As the first brave bands arrived in an area where more archaic humans lived, or where no humans lived at all, was there peace and enough food for all? And then as the centuries passed, and ever more "modern" humans arrived or filled the land with their high birth rate, was there still peace? When the game became scarcer and more aware of the danger posed by the new arrivals, did conflict arise among the distinct species of humans?

One of the great surprises of contemporary archaeology is how much of an impact humans can have on an ecosystem that had evolved without them. We seem to have an ingrained notion about "the noble savage," an idea that humans can live in harmony with their environment in a "native" state. What this view overlooks is how large an animal we are compared with the majority of creatures on Earth. Of the 5 to 30 million species on Earth today, only two or three hundred are larger than *Homo sapiens*. We are big. We eat a lot. We perturb habitat. Everywhere we go, we leave a mark.

Jonathan Kingdon, a professor of zoology at Oxford, has written eloquently about how the first human introductions into ecosystems caused substantial change. He notes that our species evolved in an ecosystem that was a heterogeneous mosaic of changing conditions and resources; we are a species adapted for change, not stability. Although much is made of our tropical birthplace on the wide savannas of eastern Africa, Kingdon characterizes that birthplace as an unstable area where climate varied and our major food resources—other mammals—fluctuated in population. Perhaps much of our nature can be understood in the context of fluctuating food resources and enormous climatic instability. We had our beginning in the interface between forests and grasslands, in the ever-changing thickets and rivers and rain shadows of a heterogeneous savanna. Life was always on the edge in this African Eden, and sometimes, perhaps quite often, conditions became so hostile that it was time to move.

In the early years of our species, we apparently fit into the African ecological mix as rare members of a poorly defended species. But with the acquisition of tools, we changed and our population increased. We became predators instead of prey, and we became numerous. We became instigators of change as well as survivors of it.

Two acquisitions by humans surely began to perturb ecosystems in important ways. First was our mastery of fire. This asset alone makes us unique in the history of the world. Yet how often did that mastery go astray, unleashing devastating forest or grassland fires? And how often did we intentionally set fire to the countryside? New information from diverse localities suggests that humans may have used fire as a weapon against nature, perhaps gleefully. We now know that both the great North and South Islands of New Zealand were essentially burned to the ground by the earliest humans to arrive there. Perhaps over the long course of human history, it is arson that has most transformed the world.

The mastery of hunting techniques may have been devastating as well. Preferential culling of young members of a prey species by human hunting would surely have affected the target species, particularly those characterized by slow breeding. If these prey were migratory, moving in and out of human territories, our impact may have been minimal. But species bound to particular territories, such deer and moose adapted to certain river valleys, would have been devastated by the introduction of humans. Extinction takes place at the local population level as well as at the species level; we must have eliminated many populations of larger mammals as our African numbers swelled.

Kingdon suggests that the impact of hominids on African savanna ecosystems can be divided into two periods. The first coincides with the period of *Homo erectus* and its direct descendant Heidelberg humans, a time of low hominid population numbers. The Erects, as the anthropologists have nicknamed *Homo erectus*, were anatomically quite distinct from us and were surely a different species from all humans on Earth today. The Heidelberg forms (perhaps an intermediate evolutionary step) were distinct from both Erects and Moderns. They were not "modern" *Homo sapiens*, and we will never know whether they should be referred to humans or how that term should really be defined. Suffice it to say that they had smaller brains than we have, smaller population numbers than the first Moderns, and very limited technology. They probably made some use of fire and had a limited repertory of hunting techniques. We do not know whether they were capable of hunting and killing larger mammals; they may have specialized in smaller mammals and the young of larger forms. During this phase, very few species would have died out as a result of the introduction of these new hominids, and those that did would have been highly specialized forms incapable of migrating to escape trouble or changing their habits enough to shield themselves from the human-like predators.

With the evolution of Moderns, however, this situation must have changed. New and better hunting techniques arose, and more widespread use of fire that devastated and perturbed vegetation patterns. That African trees today are largely fire-tolerant may be a consequence of 100,000 years of human-caused range fires. Our fascination with fire, our need to play with matches—this trait may have transformed the world in far more significant ways than we have previously acknowledged. For example, when the first Europeans took their long, covered-wagon pilgrimages across North America into the great West, they assumed they were crossing a virgin land inhabited only by vast herds of game and a few bands of "wild Indians." In reality, they crossed a land decimated of game, a land long ago radically changed by the presence of humans, where even the vegetation had been transformed by our species's penchant for fire and taste for meat.

. . . .

Our species left its ancestral homeland, Africa. But is much of our African heritage still locked within us? Our genus, *Homo*, originated in Africa and remained there for almost 2 million years. Such a long stretch of time has surely made us more African than anything else. When we left Africa, we took with us the knowledge of how to hunt large animals. That knowledge, and the habit of using it, have never left us.

Subtle aspects of human behavior sometimes show how we are still wedded to Africa. The best professor of my university days, Gordon Orians of the Unitersity of Washington, has recently completed a fascinating study showing that our African heritage still looms large in all of us.

Orians asked people at disparate places around the globe to indicate their landscape preferences among a series of pictures. To his surprise, he found that regardless of cultural identity, a vast majority of humans prefer landscapes consisting of flat grass, with a few flat-topped trees, to any other scene. Even the stereotypic America suburban landscape (house with grass in front and perhaps one tree) is distinctly African. And where do we put the tree? Never in front of the door or a viewing window; we need to see out. We like a cave or protection at our backs, some tall tree in front but not right in front, and an unimpeded view of the street—or savanna—before us. It is important to see where the predators may approach from. Orians points out that the most successful landscape architects the world over keep recreating a scene reminiscent of our east African birthplace.

Sound far-fetched? Coincidental? Orians tried another tack. He examined one hundred landscape paintings of the last three centuries. Each painting was chosen because it represented either a dawn or a sunset scene, and in each painting the position of humans within the landscape was analyzed. In the dawn paintings, humans could be found almost anywhere in the picture. In the sunset paintings, however, humans were always either next to a building or next to a tree. They were never pictured out in the open as the sun set. Sunset is when the predators begin to feed.

In the final part of his fascinating study, Orians asked people in many different geographical settings to choose their favorite type of tree from a series of drawings. The favorite trees were flat-topped, the type of tree found on the African savannas. The least favorite trees were those whose branches were so high that they could not be reached easily—the type of tree that would be no help if a lion were charging.

We left Africa but took part of it with us. Things as subtle as landscape preference may have an African origin. Our instinct for hunting large mammals, including elephants, may also be part of this African heritage.

. . . .

The first stop in our "Out of Africa" saga was the Middle East. The Moderns found a land filled with trees and already inhabited by other types of humans. In the Middle East, the Moderns encountered the Neanderthals.

Were they the same species as us? Could the two forms interbreed? That we cannot yet say.

The Moderns and Neanderthals overlapped in age (and perhaps even coexisted) for only about 5000 years; then the Neanderthals went extinct.

Were we Moderns simply better hunters who outcompeted these more archaic forms? Or did we exterminate them? Is *genocide* a fair term to describe our interaction with the Neanderthals? Again, we cannot know. Perhaps some things are best left unknown.

By 35,000 years ago the Moderns had invaded Europe. The people called Cro-Magnons were finally displacing the last of the Neanderthals there, perhaps hastening their extinction as well. With that land inhabited, we set our sights eastward. To the north lay the glaciers, to the west the great ocean. To the east, however, lay an eternity of land, much of it covered by grass. It was not prime human habitat: It was cold. But it had a resource that must have made it irresistible. The grasslands of Siberia were a haven for big game, especially the mammoth and mastodon.

By 35,000 years ago, Southeast Asia also had long since been colonized, and *Homo erectus* was extinct there. By human hands? Perhaps, but we will probably never know. Ecologists believe that most extinction occurs because of competition, not predation. Perhaps, the modern humans were superior hunters rather than wholesale murderers. Are we really Cain, or Abel, or a little of both?

The Moderns conquered the world bit by bit. They arrived in each new region slowly yet inexorably. It didn't happen in a century. It didn't parallel the taming of North America by Europeans, when several centuries saw the transformation of the giant continent's native vegetation into agriculture and concrete. Instead, the Moderns spread slowly over the globe. Even the island continent of Australia was the habitat of *Homo sapiens* by 35,000 years ago. Northern Asia, however, remained undiscovered. And beyond Asia, an even bigger territory—North and South America—still had not felt a human footfall.

The first people to reach the vast tract of what is now Siberia were Paleolithic big-game hunters. They arrived as much as 30,000 years ago, having already established a tradition of existing in this harsh climate. Eastern Siberian stone tools show some differences from what prevailed in Europe at the time and were clearly influenced by the flake cultures of Southeast Asia. Yet the major technology—the construction of large spearpoints—was formulated for killing large animals.

Cave drawings, found in France, of mammoths that lived 25,000 years ago.

The first humans arrived in Siberia in a time of slight warming, and this warmer period may have encouraged the spread of humans into an otherwise hostile region. Yet soon after their arrival in Siberia, the earth began to cool again, and by 25,000 years ago a major glacial event was well under way. In Western Europe and North America, the mile-thick continental ice sheets were inexorably spreading downward over vast regions. In Siberia, however, there was so little moisture that the ice could not form. Into this treeless, frozen territory, humans expanded ever eastward. Because there was so little wood, the hides and antlers of their prey became important resources, and the very bones of the largest and perhaps principal quarry—mastodons and mammoths—were used for housing.

As humanity crossed Asia and settled in Beringia some 18,000 years ago, the continental ice sheets covering large portions of North America reached their maximum extent and thickness and then, slowly, began to melt. As they did so, the level of the sea began to rise. As late as 14,000 years ago, the continental glaciers covering most of Canada and large portions of what is now the United States were still melting under gradually rising temperatures. Soon thereafter, however, a new event accelerated the melting process. When enough ice had melted so that the glaciers no longer extended out to sea from the coast, the calving of icebergs from the eastern and western coastlines of what is now Canada and the northern portions of the United States could no longer occur. Each spring during the period of maximum glaciation (about 18,000 to 14,000 years ago), great fleets of icebergs were launched into the coastal oceans. This kept the waters cool and created very cold winds that cooled the lands as well. When icebergs stopped forming, warmer onshore winds arose, and the ice began to melt in earnest everywhere on these continents.

So much water had to go someplace. It began its return to the sea but in large areas was barred by topography, for as the ice melted, the land on which it had so long rested began to rise. The great weight of ice had compressed the land, and with its removal began a glacial rebound that continues to this day. Still the ice melted, forming sea-sized freshwater lakes in the middle of the land mass.

The melting fronts of the glaciers must have been extraordinarily harsh places. Incessant, strong winds characterized the retreating glacial walls. So strong was the wind that it created great piles of sand and silt, a sediment called loess. The winds also carried in seeds, so the drifting soils in front of the glaciers were soon colonized by pioneering plants. First came the ferns, then more complex plants. Willow, juniper, poplar, and a variety of shrubs were the first stable communities to transform this ancient glacial regime, and soon successive communities of plants arrived. In the more temperate west, low forests dominated by spruce were the norm; in the middle, colder parts of the continent, permafrost and tundra. Yet everywhere the glaciers were in retreat, and as they melted northward, they were pursued by a front of advancing tundra, itself soon followed by vast spruce forests.

Yet *forests* may be the wrong word to characterize the vegetation. The spruce communities were really more open woodland than dense forest, consisting of copses of trees interspersed with grass and shrubs. By no means were they similar to the great, thick Douglas fir communities found in the few remaining old-growth forests of the northwest, places where dense underbrush and fallen rotting logs make passage by large game—or humans—exceedingly difficult.

South of the ice in North America, throughout the Ice Age, a variety of habitats existed. There were forest, tundra, grassland, and deserts, and there were plants sufficient to sustain enormous herds of giant mammals.

Mammoths and mastodons first reached North America from Asia about 1.5 million years ago with the arrival of a beast called *Mammuthus meridionalis*. Other proboscideans related to elephants were already here, such as the strange, pointed-tusked gomphotheres. They arrived by the same path that many other mammals, including humans, followed: the land bridge connecting eastern Asia and North America. With the lowered sea level of the Ice Age, great expanses of continental shelf that are now far under water were dry land, and across these corridors many large and small animals migrated. Yet this bridge was not a constant feature; it appeared and disappeared in concert with the glacial advances and retreats that controlled the level of the sea. It was as if a drawbridge went up and down, each short period of connection letting in some new group of mammals.

The isolated populations reaching North America soon underwent their own evolutionary changes, and by 18,000 years ago a diverse and unique assemblage of giant mammals lived in both North and South America. Then, about 12,000 years ago, an awesome new predator arrived.

. . . .

By the time humans massed at the jumping-off place for the New World—the last two great continents to be colonized by the bipedal apes from Africa—there was but one species of hominid left. Even 20,000 years ago, all else akin to us was long extinct; the Erects had been dead for a hundred millennia, the Neanderthals for tens of millennia. By this time we had mastered the game of evolution. The other species still did not know, as they cannot know today, that the game is over. Today, species after species disappear in the most widespread mass extinction that Earth has seen since the days of the Chicxulub comet some 65 million years ago. We are the comet now. And not only have we won the game of evolution; we control the rules of the game.

Nineteenth century reconstruction of Mammalian fauna (ground sloths and mastodon) encountered by the first humans to arrive in North America. (From L. Figuier, La Terre Avant Le Déluge, 1864.)

6

The Hunger

IMAGINE YOU ARE A HUNTER newly arrived in North America following a long, arduous walk from Arctic regions nearly 12,000 years ago. You come armed, for your people have hunted big game for a thousand generations, and in all that time they have never once raised a crop. Your people lived for meat, for liver and heart and tongue, for the red blood that sustained us all. Other food occasionally sufficed, such as roots and berries and other edible plant parts. But even these were rarities in the cold far north from whence you came. You came as hunters, and you arrived in a hunter's paradise, a place never before seen by human eyes—at least that is what most scientists believe.

There is continuing debate about when the *first* humans arrived in North America. We know they did so by traversing a land bridge hundreds of miles wide known as Beringia, which, because of the much lower sea level at that time, connected Asia and North America. Most archaeologists agree that the oldest reliably dated human artifacts yet found in North America belong to a people we call the Clovis (for the name of the small town in New Mexico near where their artifacts were first unearthed). The Clovis artifacts, however, are not limited to this locality: They have been found at sites throughout much of North and South America.

The Clovis people appear to have entered North America from Siberia between 12,000 and 11,000 years ago, a time that coincides with the retreat of North America's great glacial cover. They found a continent filled with wildlife, much of which could be characterized as "big game," such as elephants and rhinos and bears. Yet within about 1000 to 2000 years after their arrival, most of this game was extinct, and the creatures that did survive, such as deer, wolves, and foxes, are hardly candidates for a big-game zoo. This great extinction—truly a mass extinction—represents one of paleontology's most fundamental mysteries. The problem is made even more vexing by its nearness to our own time: The cause of a mass extinction so devastating *and* so recent in geological terms should be easily identifiable. Yet such is not the case. Two suspects are considered the most likely culprits. Was it climate change or the actions of the Clovis people themselves that killed off the Ice Age megafauna?

But were the Clovis really the first people in North America? Many tantalizing hints from tenuous archaeological data suggest that an earlier people, perhaps as far back as 40,000 years ago, were the true first human pioneers on this continent. Radiocarbon dating of material associated with human artifacts from the Meadowcroft Rockshelter site in Pennsylvania has yielded ages between 19,000 and 14,000 years. Another site has been found in Chile that may be even older than the Pennsylvania site, indicating that humans reached not only North America but also South America long before the first evidence of the Clovis people. This view was further strengthened by a discovery announced in April 1996, when a cave in Brazil yielded well-dated fragments of bone and tools from people who lived contemporaneously with the oldest of the Clovis—about 12,000 years ago—but seem to have had none of the implements and weapons we associate with the big-game hunting that the Clovis apparently engaged in. The new South American record shows a people living in a jungle region devoid of big game, subsisting on berries, fruit, and other items, just as many local peoples do today.

The scientific jury is still out on this issue. But on one subject most archaeologists agree: If the cultural objects associated with these early finds did indeed come from a pre-Clovis culture, it appears *not* to have been involved in big-game hunting. There is no evidence of the production or use of weaponry capable of assaulting the large mammals of North America. By contrast, the Clovis produced large, 6-inch stone spearpoints that were very beautiful—and surely very effective. Such weaponry had but one purpose: to bring down big game.

We know the route the Clovis followed to get to North America. As the glaciers began to recede from the northern tier of the Americas roughly 15,000 years ago, huge regions that had long been covered with thick glacial ice emerged. The great glacier covering what is now Alaska, British Columbia, Alberta, and the Yukon separated into a western and an eastern lobe, with a wide, ice-free corridor down the middle: a slow road south, beckoning brave hunters to journey into the unknown. At the southern end of this long, wide valley lay the Great Plains of what is now Canada and the United States, areas we call Alberta, Montana, Wyoming, the Dakotas, Nebraska, and Kansas: huge, productive grasslands that must have been teeming with game.

Why did the first Americans move south? Was it simply the harshness of the land in Beringia, the nether reaches of this great ice-free corridor and the land bridge to Siberia? Was population pressure from new migrants the cause, displacing more and more people southward? Did their folklore include legends of warm lands to the south, somewhat like those legends that enticed sailors in the time of Columbus to seek a great land to the west? Or was the reason for these marches more mundane, such as the need to find new hunt-

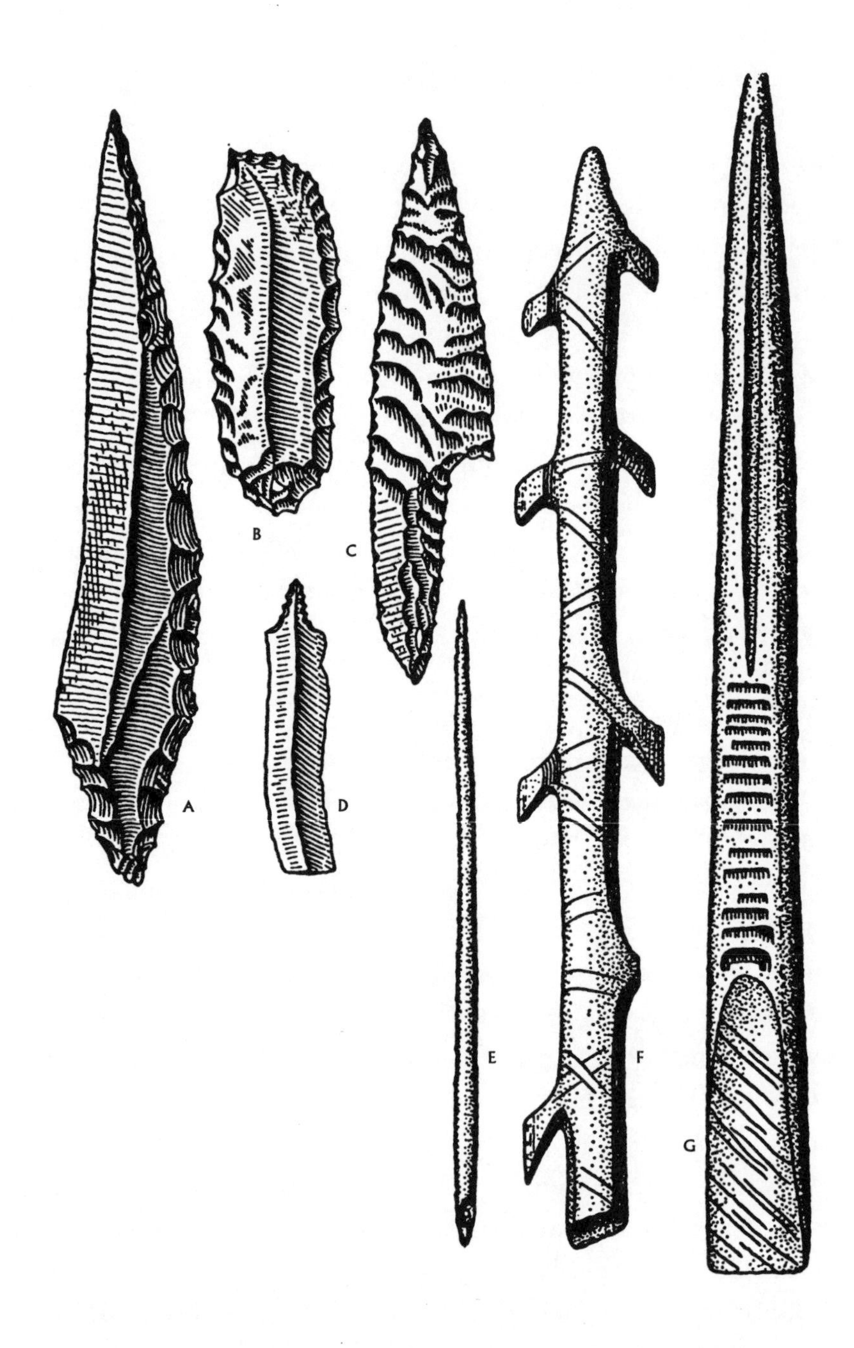

Stone and bone artifacts from Stone Age European localities. Implements and weapons (actual size) of the Ice Age (Stone Age) hunters. Made of flint: A. Pick or knife B. Scraper C. Serrated pick D. Boring tool. Made of bone or horn: E. Needle with eye F. Harpoon G. Spear point (engraved).

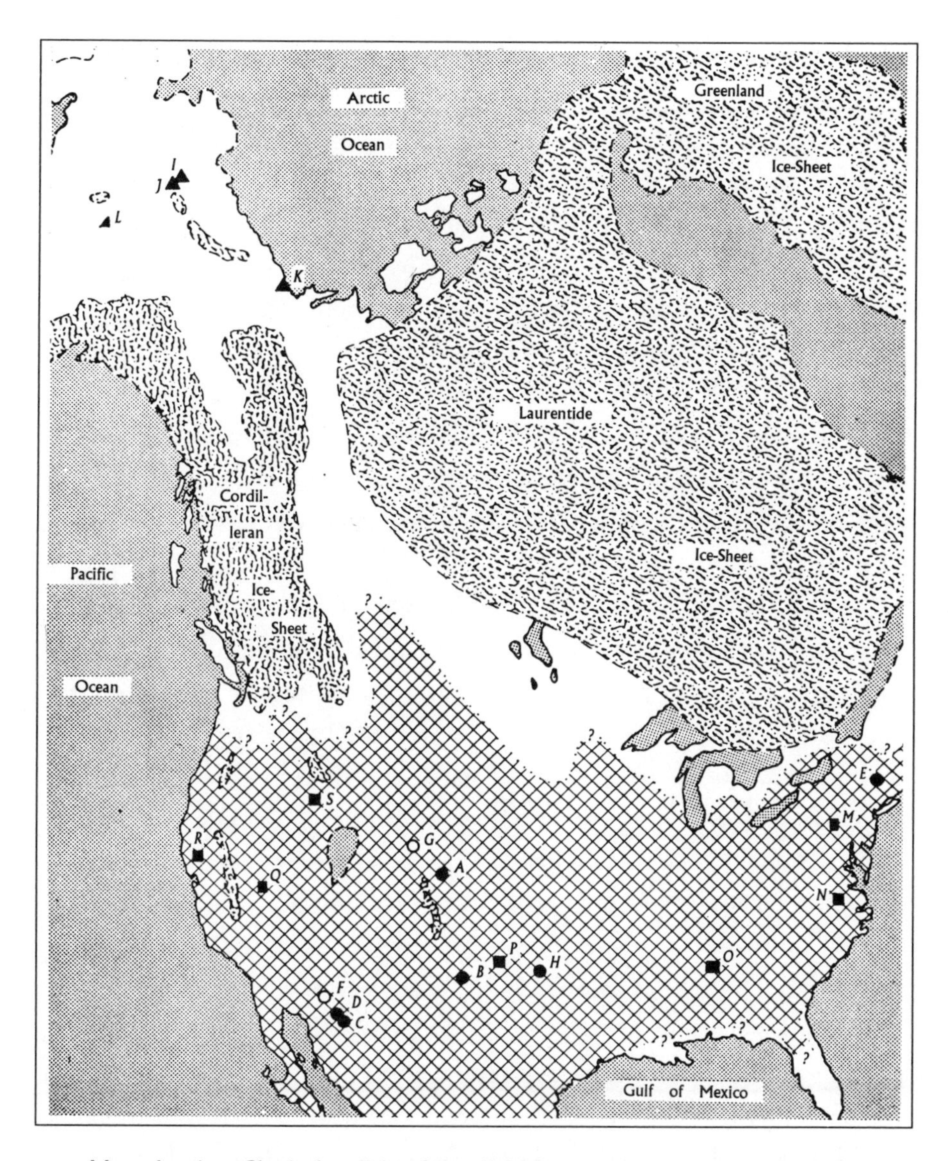

Map showing Clovis localities from 11,000 years ago, superimposed on a slightly older time when the land bridge between Asia and North America was still emergent. (Modified from Science Magazine, C. Haynes, 1964; with permission.)

ing grounds? These people were not farmers. They lived in a land that offered very little food from plant resources. They needed fresh meat to survive.

We can imagine these treks. They were probably not well-thought-out affairs; Stone-Age people then living in Asia did not set out to find the

New World for economic gain, as later explorers would. Their starting point, the land bridge joining Asia and North America, contains archaeological evidence of human habitation as much as 25,000 years ago and perhaps even 50,000 years ago. Yet for many millennia the way south was blocked by ice. Starting 15,000 years ago, that way began to open. By 12,000 years ago, the great highway was available for any humans brave or desperate enough to travel ever south, a gradually warming sun in their eyes.

The mettle of the people who lived in this far northern clime had to have been forged in hardship; to survive they must have been skilled hunters. Although there is abundant evidence that herds of game lived alongside the humans in Beringia, the quarry must have been well acquainted with humans and their hunting skills. And the harsh climate required clothing, fire, and shelter.

When they finally arrived on the Great Plains, the earliest pioneers in North America prospered, multiplied, and continued to travel. Travel may have been their way of life, for without agriculture they were not tied to any plot of land. Thus they may have followed migratory game, or they may have been forced to move when game in any region was exterminated by their hunting. These people, whom we now call the Clovis (although they were actually many small bands, not a single tribe), appear to have undertaken great and rapid migrations; often the artifacts we have found in sites they inhabited are made of stone or other material that must have originated hundreds or even thousands of miles away.

We know very little about these people. We know of no single name or face, nothing about their social orders, nothing about how they lived or how they worshipped or who their great leaders were. Two things are undisputed, however. These people knew how to kill big game, and they knew how to travel. They did not rest when they arrived in North America, for within a millennium of conquering North America, they had reached the southernmost tip of South America as well.

The first discovery of what we now call Clovis artifacts came in the 1920s, when a cowboy noticed large bones protruding from an eroded channel bank in New Mexico, near the Four Corners region. Paleontologists eventually came to this site, near the town of Folsom, to find the fossils of an extinct species of giant bison. Although bison fossils of this age were not unknown here, the objects found among these bones were rare and new: stone tools and weapons, indicating that the creature these bones belonged to had died by human hand.

The find sent shocks through the archaeological establishment. In the 1920s, the prevailing wisdom dictated a very late arrival of humans in North America, certainly well after the giant bison were thought to have gone extinct. But unless someone had carried out a very elaborate prank, here was evidence showing not only that humans had coexisted with at least one species of now-extinct mammal but also that they had hunted it.

Did these ancient people hunt even larger denizens of the Ice Age world as well? Bison hunting, after all, does not seem so extraordinary. In the 1920s, recollections of the great bison hunters of the previous century still loomed large in the memory of American culture; that ancient humans had done the same was probably not a great surprise. It was also argued that perhaps the extinct bison were not all that old (radiocarbon dating had not yet been discovered in the 1920s), and the actual age of these bones, in thousands of years, could not be ascertained. In the early 1930s, however, such questions became moot.

In 1932, remains of several mammoths became exposed following a flood along the North Platte River near the town of Dent, Colorado. The mammoth bones were excavated, and as in the bison find, human stone tools were found among the bones. Unlike the bison find of the 1920s, however, this site contained far more than a single dead bison. It was discovered during the excavation that the Dent site was the location of an ancient pond, a place where many large mammals came to feed and drink—and where they were hunted. The region seems to have been occupied or visited by humans for at least 2000 years, and it was eventually found to be older than the bison site.

The artifacts found at the Dent site were numerous and extraordinary. The most characteristic were large, fluted projectile points. These spear tips were usually at least several inches long and were crafted out of hard types of rock such as chert or the quartz-rich minerals. Sometimes they were even sculpted from quartz crystals. These artifacts matched others previously found near Clovis, New Mexico, and thus came to be called Clovis points.

Since the first finds near Dent, similar projectile points as well as other types of tools have been found over wide geographical regions of North America. The similarity of these Clovis points to stone tools and weapons found in archaeological sites in Europe and Asia suggests that the Clovis tradition of making large spearpoints originated in the Old World. (Another possibility is that it developed independently in the two regions.) In any case, it is clear that humans 10,000 years ago had the capability to be big-game hunters. And judging by the types of animal bones associated with the Clovis points in North America, mammoths and mastodons may have been a favorite North American food source.

Did the Clovis people actually hunt down healthy adult mammoths, mastodons, camels, and horses, using their stone implements to kill these largest of land creatures? Or were the Clovis people more opportunistic, herding the lame and young into pools of water, or sending whole herds over cliffs, and then using their stone tools not for killing but for butchering and removing the meat? Were these people the great mammoth hunters that Jean Auel and other writers of popular fiction seem to imagine, or were they mammoth scavengers?

Clearly, killing a mammoth while armed only with a stone-tipped spear (or even many stone-tipped spears) must have been no easy endeavor. How would you kill a charging mammoth or mastodon with a spear? Head shots won't do; a spear thrown at the head of an elephant would never get through the thick bones of the skull. It looks like the only hope would have been a thrust through the rib cage, into the heart and lungs. Yet this must have been no mean feat, for a variety of reasons.

First, you have to contend with the hide itself. Archaeologist George Frisson has calculated that a mammoth's hide would be about half an inch thick. This hide must have been tough, and surely there was a thick insulating layer of fat beneath it, so that the creature could withstand the cold winter conditions near the glaciers. To penetrate into the vital parts of the mammoth or mastodon body cavity, the spear tip had to be long, sharp, and sturdy enough not to snap on impact.

Archaeological studies and experiments indicate that two means of penetration were utilized: conventional spears, in which a point securely attached to a straight shaft is either pushed or thrown into the body of the prey, and the use of an atlatl, or spear thrower, in which a piece of wood held by the hunter was used to launch a short-handled shaft with a great force and velocity.

In both cases, the most important part of the whole weapon was the tip. The end of the spearhead had to be pointed enough to allow initial penetration, thin enough along the length to allow continued penetration, and strong enough so that the tip did not snap off at first impact. The material used for the Clovis spearheads, as well as the workmanship necessary to create them, had to be of the highest quality. Faulty material or workmanship would be costly indeed.

The points that have survived from the Clovis culture are extraordinary, exquisite works of art, aesthetic and sobering at the same time. They were made by percussion flaking, and the master craftsmen who made the points sought out stone that had high flaking quality and sturdiness under tension. Quartz-rich stones were clearly the best materials, because quartz minerals produce curving fractures and thus can be shaped, through hundreds or thousands of careful blows, by the sculpting of the master chippers.

Many fascinating questions remain unanswered. The most important is whether the Clovis people hunted mammoths or scavenged mammoths. Recently, the hunting capability of what was assumed to be the greatest hunter of all time, the dinosaur *Tyrannosaurus rex*, has come into question. Paleontologist Jack Horner has had the temerity to suggest that old *T. rex* was more scavenger than hunter—that the wily giant waited until some smaller and more agile predatory dinosaur made a kill and then sauntered over and stole the prey. In analogous fashion, it has been suggested that perhaps the Clovis people did not hunt mammoths and mastodons at all but simply killed wounded or water-trapped animals or stole the remains of freshly killed juvenile or old elephants from other predators.

There is not much evidence to test these hypotheses. There are a little more than a dozen sites in North America where Clovis artifacts—most importantly, of course, spearpoints—have been found associated with elephant bones. Of these, only one site, the Naco site in Arizona, has yielded hard evidence that the fossil elephant found in association with human artifacts was actually killed by the Clovis. At Naco a fossil mammoth has eight Clovis spearpoints within the skeleton.

The report on the Naco kill site, published in 1953 in the journal *American Antiquity*, is riveting reading, as much for how it was written as for what it says. After describing how a fossil mammoth bone concentration was discovered by farmer Marc Navarette and his father, Fred, the author of this article, Emil Haury, goes on the describe the subsequent excavation by archaeologists. Navarette *pere et fils* are lauded in the third paragraph of the article for their "exemplary attitude and alertness," which "shine as a beacon on the relationship between the interested amateur and the specialist." Then we learn that the specialist managed to excavate the entire skeleton in four days. Either Dr. Haury was the most proficient excavator in the history of paleontology, or this was a scrape and dash operation. Nevertheless, this most important of North American archaeological finds yielded tantalizing information. Of the eight large spearpoints found among the bones, one was near the base of the skull, another in the neck vertebrae, one in the shoulder area, and the rest among the ribs. The tips were made of fine chert, a hard, quartz-like material. Haury concluded his article in straightforward fashion:

> Prior to 10,000 years ago . . . a Columbian mammoth was killed by hunters who hurled no less than eight stone tipped spears into it. The animal fell on a sand bar adjoining a stream and what remained of the carcass after the hunters had salvaged the parts they wanted was soon covered by a succession of deposits which register the subsequent climatic history of the region. Though variable in size, the spearpoints were all of one kind, called by students of early man the Clovis Fluted type. These are almost identical with others found to the east and northeast in the Plains, thereby extending the known distribution well to the west.

Haury clearly had a time machine: His find is the most persuasive evidence that the Clovis people were hunters, not scavengers.

Scavengers or hunters? The evidence suggests that hunting was predominant. The Naco site offers the best evidence of actual hunting, but several other sites also have associations of skeletons and spearpoints. Some of these suggest that the hunters may have preyed not on adults but on the younger mammoths. At a ranch near Naco, for instance, the fossils of 13 young mammoths were found in association with spearpoints and fire pits. In Colorado, another site contained many juvenile mammoths in association with stone tools. Surely a young mammoth would be easier prey than an adult, though if mammoths showed any of the social structure seen in living elephants, the culling of younger members of a herd must have incurred the wrath of the mother.

Not all sites indicate that the mammoths were killed by spears. A site

Mammoth excavation site, Nevada, with excavator Dr. Stephanie Livingston, a former student of Dr. Don Grayson. Note the head and tusks with other bones in foreground. (Photo by Don Grayson.)

near Dent, Colorado, is in a landscape that suggests the mammoths were stampeded over the edge of a bluff and later butchered with the stone tools. Surely not all such sites are human kill sites. Yet the tool associations suggest that humans were involved in more than a few mastodon and mammoth deaths in North America.

. . . .

What do we know of the mammoths and mastodons? They were anatomically very similar to living elephants. Paleontologists seem to agree that mammoths, mastodons, and living elephants were probably very similar in behavior and biology as well.

Although we know little about mammoths and mastodons, there is an enormous literature on the biology of modern elephants. They are social; their herds exhibit a clear hierarchy of dominance that supports survivorship; herds are made up of well-organized family groups. Elephant herds in Africa can include more than 30 members. Social behavior within the herds is complex, for these are highly intelligent animals. The herds are organized in familial and kinship groups made up of adult females and their offspring. Elephants are not solitary beasts.

The same can probably be said for both mastodons and mammoths. Yet many accounts of mastodon biology, and to a lesser extent mammoth biology as well, propose that herd behavior was less important, or even nonexistent, in these extinct taxa. This interpretation derives in large part from most mastodon finds being of single specimens, which gives the impression that they were solitary. There are, however, compelling reasons to believe that both mammoths and mastodons belonged to herds. This inference comes from biological parameters pertaining to growth.

The most important parameters for reconstructing the lives and biology of the mammoths and mastodons appear to be the following:

1. The relationship between body size and life history parameters, such as gestation period, duration of lactation, age at first reproduction, the time spent feeding daily, and life span. By studying body size in modern ele-

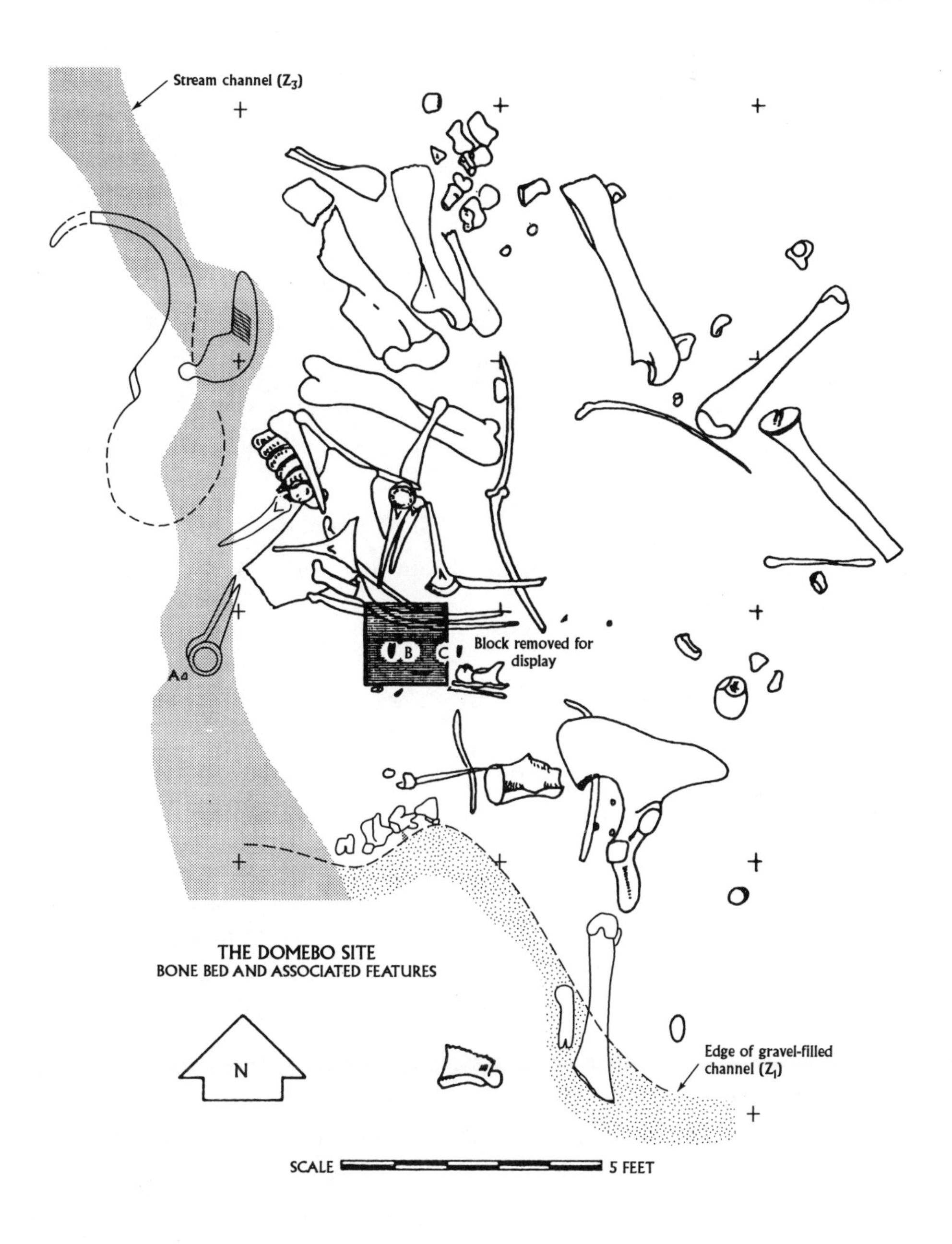

Diagram from Domebo site, North America, showing the position of scattered mammoth bones as they were found. Articulated skeletons in a life position are rarely recovered at bone sites.

phants and comparing these sizes to the important natural history traits listed above, we can make estimates about the same traits in the extinct species. For instance, on the basis of body size alone, we can estimate that the gestation period (the time between fertilization and birth) was longer in mammoths and mastodons than it is in living elephants.

2. The anatomy of the digestive system and its effects on feeding strategies and foraging behavior. We believe that all three groups had similar digestive systems and thus needed to eat the same quantities of food.

3. The diversity and distribution of foods that can be utilized. By knowing this about elephants, we can make guesses about the types of foods eaten by the extinct forms.

4. Tolerance of water shortage. This is extremely important. Drought is a leading cause of death in elephants and was probably just as dangerous to Ice Age species.

5. Liability to predation. Healthy adult elephants have no natural predators other than humans. The same was probably true during the Ice Age.

In all of these parameters there was probably marked similarity among mammoths, mastodons, and elephants. Differences must have been present as well; no species is exactly the same as any other. And although these differences may have been relatively minor, even minor differences can be crucial in deciding which species survives and which goes extinct. For instance, mammoths and mastodons were heavier than modern elephants are. This conclusion is based on comparative studies of bone sizes. The modern elephants are more "gracile," they have thinner bones for similar lengths of bone. The larger mammoths and mastodons would thus have been slower afoot, would have required more food, and would have had longer gestation periods, yet they would have been capable of less extensive migration in search of food or water. All of these factors could have made them more vulnerable to extinction.

Mastodons required 10 years to develop from birth to maturity and, like modern elephants (and humans), surely needed maternal help and protection for extended periods of time following birth. This behavior is the source of the matriarchal family grouping that in elephants creates herds;

mastodons, with a similar life history strategy, must have been herd animals as well. All of this raises an ominous point: Mammoth and mastodon populations would have recovered more slowly—perhaps far more slowly—than modern elephants following a period of population stress.

Another aspect of elephant life history is pertinent to mastodons and mammoths and may be a clue to their extinction. Studies on elephant herds that have undergone environmental stress for some time show that such herds exhibit a marked change in age classes. When facing long periods of drought, elephants effectively "sacrifice" their young. The mothers even fend the young away from water holes so that the adults can drink. Under such conditions, the young quickly die. This harsh behavior enhances the survival of the reproductively viable members of the herd and thus can enable the herd to return to a normal population age distribution—and survival—when and if more favorable environmental conditions return or are restored after migration. Nevertheless, extended periods of stress lead to an aging of the herd, and when the population of young remains abnormally low for extended periods of time, the herd is doomed. Healthy populations of elephants show subadult populations of 30% to 50% of the total herd number.

Just such a situation may have occurred in the Ice Age, only in this case both environmental stress and human hunting may have severely affected the age classes of the mammoths and mastodons. There can be little doubt that human hunters would have focused their hunting activities on the smaller elephants. Smaller individuals were far less dangerous and would still yield an extraordinary quantity of meat.

Studies of modern-day elephants suggest an important lesson: Because of their size, intelligence, and ability to tolerate a wide variety of environmental changes, elephants are remarkably resistant to extinction. African elephants can be found from the edges of deserts to the deepest rainforest; they live in the savannas of the plains as well as the highlands; they span the great continent of Africa. A range this extensive means it would take extraordinary circumstances to kill off a species of elephant. Might this not have been true in the past as well?

The extinctions of the mammoth and mastodon in North America, and of the great elephant-like gomphotheres in South America, were completed by about 9000 years ago at the latest. Soon thereafter, the Clovis people—strongly associated with and perhaps responsible for these extinctions—themselves disappeared from the face of the earth. And they vanished not through extinction but through cultural change.

. . . .

Between 11,000 and 9000 years ago, there seems to have been but a single Paleo-Indian culture, the Clovis culture, throughout North and (later) South America. During this period great climatic, geographical, and vegetational changes were taking place in North America. Large regions that had been under ice or covered by giant lakes emerged as dry land, at first barren but eventually colonized by rapidly changing plant communities.

Beginning about 9000 years ago, a great diversification of Paleo-Indian cultures began, culminating in the many tribes of Indians found in North America today. In essence, the Clovis themselves became extinct as they evolved into the great diversity of Native Americans. That this change from one to many cultures occurred 9000 years ago is probably very significant. Radiocarbon dates of kill sites show that by 9000 years ago, only slain bison are found, not slain proboscideans. After that time the Clovis sites ceased to exist—as did the proboscideans. The great mammoths, mastodons, and gomphotheres had entered the oblivion of extinction, and the Clovis culture disappeared with them.

Was the Clovis culture dependent on the proboscideans? One of the mysteries is that so few other large mammals show up in the Clovis kill sites. Surely there were easier animals to bring down. We know nothing of mastodon and mammoth social organization, but all modern elephants show strong herd structure, including a matriarch and well-organized family groups. This behavior characterizes both the African elephants, *Loxodonta*, and the Asian elephants, *Elephas*. Hunting and killing members of elephant herds had to be a very risky business. Why single out elephants when much other game is present, unless you have some very good reasons for doing so?

And how important were the elephants to Clovis religion and lore? The Clovis have left us no cave paintings, no scrap of writing, no indication of how they felt about the great beasts. Were they loved? Hated? How many Clovis husbands and sons were killed or maimed by elephants? And when the last elephant was gone, what then? How was this explained around the fires, in stories about the old days when the elephant herds dotted the plains? How did the Clovis rationalize the disappearance of the great elephants?

The surviving mammals flourished. White-tailed deer spread across North America, while mule deer ranged through the high plains and western mountains. Huge herds of pronghorn antelope spread across the western prairies, sharing the bountiful grass with herds of plains bison and elk. Prairie dog colonies covered great expanses of territory, and packs of wolves, mountain lions, and great bears vied for food among the abundant herbivores of the west. In the eastern regions, an even larger bison lived in the vast forests along with herds of elk and giant moose. Great flights of waterfowl and other birds blackened the skies. It was a continent rich in animal life but very poor in its *diversity* of large mammals.

Humans had to change as well. Their technology changed, for the spearpoints show an evolution. Great "elephant gun" spearpoints were no longer necessary; other than the bison herds, the biggest game was gone. How did the Clovis react to this? Providentially, in early 1996 a new window into this past emerged. Scientists at the Nevada State Museum were able to employ sophisticated new radiometric dating techniques to re-examine a mummy held in their museum since 1940. Known as Spirit Cave man, this ancient corpse was found in a Nevada cave in 1940 and had been thought to be about 2000 years old. To the scientists' surprise and delight, new tests showed the mummy to be far older—about 9000 years in age, and thus at least twice as old as the famous Ice Man mummy found in 1991 in the European Alps.

The sophistication of the artifacts found with the Spirit Cave mummy was also surprising. He was wearing moccasins and was wrapped in shrouds woven from marsh plants. The scientists even investigated the remains of his last meal by examining his mummified intestines. They found only fish bones, not large mammal meat. The re-evaluation of the Spirit Cave mummy came about because of rapidly improving techniques in one of the best-known dating procedures, carbon 14 dating. Much of the current excitement about and reinterpretation of our recent past now revolves around this all-important procedure.

. . . .

The debate over whether humans were involved in the destruction of whole *species* of giant mammals in North America has been under way for well over a century. From the beginning, the pivotal issues have been the rate at which the fauna went extinct and the reasons for those extinctions. Only two possibilities seemed plausible: The great Ice Age fauna of North America were killed either by climate change at the end of the Ice Age or by humankind. The lines among archaeologists were soon drawn, and influential voices sounded on each side of the debate. Perhaps no voice in this century was more authoritative than that of the greatest natural history essayist of all time, Loren Eiseley.

Eiseley takes an intriguing tack, in the sense that his conclusions are exactly the opposite of those drawn today from the same premises. Several of Eiseley's contemporaries argue that the Ice Age megamammals died out gradually and therefore human predation, rather than climate change, must have been the primary cause. Eiseley saw things differently:

> It is quite true that at first glance it seems odd that a fauna which had survived the great ice movement should die at its close. But die it did. The reasons are difficult to give in exactitude. Perhaps many of these forms over thousands of years had become too well adapted to glacial conditions. Time may have proved too short for readjustment to a warm climate. At all event they sickened and died. No doubt man contributed in a small, brief way to that final vanishing. But there is no evidence that he greatly altered events which were destined to transpire as they did without his interference. (Eiseley, 1943, p. 211).

Eiseley was not alone in taking this view. Ned Colbert, one of the three greatest vertebrate paleontologists of his time, took a similar stance: "Then at the end of the Ice Age when the last of the great continental glaciers was retreating to its present Arctic limits there was a relatively sudden and widespread extinction of mammoths throughout the world. Was man concerned with their extinction? It seems hardly probable." (Colbert, 1940, p. 103).

The debate continued to little effect. Specific hypotheses about the cause depended on a precise chronology of events, and no such chronology, other than the superpositional relationships of the fossils, existed. This state of affairs was largely due to the difficulties of artifact and fossil dating, for there was no reliable way to correlate or date the last known occurrences of various fossils in the glacial sediments. In the mid-1950s, however, a powerful new tool revolutionized archaeology and Ice Age paleontology: radiocarbon, or carbon-14, dating.

Comparing the fractions of the relatively rare isotope of carbon, carbon-14, to those of its far more abundant sister, carbon-12, yielded a method of actually determining the age of some organic component, such as bone or wood, became readily available. This method relies on the concept of *half-life*, the amount time required for half of the radioactive isotope to decay into other isotopes or elements. Using this new technology, many scientists began to date the last occurrence of North American Ice Age fossils. It was soon discovered that many of the last occurrences of now-extinct mammals from North America seemed to date from approximately the same period, immediately after the retreat of the last known North American ice sheet. These extinctions thus coincided with two major events: a climate change of enormous magnitude, which caused the retreat of the ice, and the arrival of the first humans in North America. Many people began to wonder again whether there was a direct link between these factors. One man, however, seized upon the "coincidence" of humankind's arrival and the great mammalian extinction with such fervor that his name will forever be linked with a most unsettling hypothesis: that we humans wiped out the great mammals through excessive hunting. The name of this idea is the overkill hypothesis, and Paul Martin of the University of Arizona was its champion.

. . . .

The term *overkill* was first used by Martin in 1967 to describe his view of the Pleistocene extinctions, but he was by no means the first scientist to propose the idea. As long ago as 1799, naturalists such as George Turner thought that American mastodons had been killed off by American Indians. Turner, of course, had no idea of the age of the mastodon bones that were found in profusion at various places in the eastern part of North America. The early evolutionist Lamarck also thought that Ice Age fossils found in Europe, belonging to species now clearly extinct, must have been killed off by humans. Other scientists of the time later agreed. In fact, by 1860 the famous biologist Richard Owen of England even wondered whether mass extinctions could have *any* cause other than humans.

Many early geologists suspected that human hunting had something—perhaps everything—to do with the disappearance of the Ice Age megafauna. But testing this assumption proved difficult. First, sites with both mammalian fossils and human artifacts were extremely rare. Second, there was no way to deduce the relative age of the fossils, so no reliable estimate could be made about rate of extinction or even the period of overlap between faunal and human occupations. But as I've mentioned, all that changed in the 1950s and 1960s as a result of two novel circumstances. First, increasing archaeological effort revealed first tens and then hundreds of new paleontological and archaeological sites. Work done at these sites greatly increased our information about the Ice Age fossil record as well as the areas occupied by humans. Second, it became common practice to date artifacts and fossils from these sites with the newly developed radiocarbon dating methods. These new data made it possible to reappraise competing hypotheses about the extinction. Paul Martin made this his life's work. Even early in his career, he had little doubt about the cause of the extinction, as shown by this statement from 1963: "Large mammals disappeared not because they lost their food supply, but because they became one." Martin's achievement, however, was not in simply restating a theory by then more than a century and a half old. Instead he revitalized the debate by presenting new data, reformulating the questions to be addressed, and making specific predictions that could be tested scientifically.

Martin's most valuable weapon was the wealth of new radiocarbon dates becoming available to archaeologists in the late 1960s. These data convinced many scientists that the main wave of extinctions had culminated about 11,000 years ago. The chronological data led to two crucial assumptions about the timing and longevity of the extinctions. *First, it was assumed that whatever had caused the extinctions of some of the animals had caused the extinction of all. Second, because no fossil remains of extinct mammals were found in any archaeological site of age 10,000 years or younger, it was assumed that all of the extinctions had been completed by that date.*

I met Martin in the late 1980s. So often, the reality of such meetings defies our preconceptions. He had pioneered such a sobering hypothesis—one based on the blood of untold large animals spilled by human predation—that I somehow expected a warrior incarnate. Instead I met a quiet, polite man walking with a cane. He spoke to us about his theory, first proposed in the 1950s and still viable in the 1990s despite almost incessant attempts, over the last three decades, to discredit it. Martin's two most powerful arguments were that the extinctions on each separate continent seem to have occurred so soon after the arrival of humans and that, unlike other extinctions from the geological past, where many diverse groups of animals and plants fell victim, only a very restricted group of animals seem to have disappeared during the Ice Age: Almost all were large mammals.

In 1967 Martin published his theory in great detail. He noted that the first humans known to have settled North America, the Clovis people, did so between 12,000 and 11,000 years ago. He also noted that by about 1000 years after this initial colonization, most of the extinctions among large North American land mammals had been completed. Martin proposed that the Clovis people rapidly hunted many species to extinction.

Martin argued that the extinctions devastated only large mammals, their predators, and the scavengers that would have been ecologically dependent on the extinct mammals. This would seem to rule out climate as the major cause of the extinctions, because climate change should affect all trophic or energy groups, not just one group of large herbivores and those dependent on them. If climate or some other agent had produced the extinctions, he argued, it should have cut a much wider swath through North America's biota, yet very little loss of invertebrates, small mammals, reptiles, or ambphibians seems to have accompanied the megamammal extinction.

To Martin, only human predation could account for the observed extinction patterns. The Clovis people, newly arrived from Asia, found a wide land empty of humans but filled with big game. With hunting skills honed by the hardships of their long Siberian habitation and their eventual trek through the cold northern wastes, and armed with exquisitely produced stone spear tips, the Clovis began to decimate the great herds of mammals. With a plentiful food supply, they quickly increased in numbers and spread across the continent. Martin called this rapid spread of the Clovis, leaving slaughtered populations and extinct species in its wake, a blitzkrieg. His blitzkrieg model envisions a mobile group of humans, well equipped and skilled in big-game hunting, passing through previously uninhabited continental areas and exterminating the big-game fauna so rapidly that few or no kill sites are left behind.

Martin certainly has attributes of the stratigraphic record on his side. Let us assume he is correct: In a period of between 1000 and 2000 years, human hunters, though small (at least initially) in numbers, move across the continent and kill off so many members of the larger-animal population that they go extinct. The remains of this butchery should thus be found only in beds spanning a thousand years of sedimentation. If the sediments were from the deep sea, then we might have a chance of detecting evidence of these events. But on terrestrial systems, where the sediment is deposited from lakes or rivers, we would expect very little net accumulation in a thousand years. Yet the killings presumably were not taking place in lakes or rivers; they were being made on prairies, woodlands, and other open land areas. These regions received very little net sedimentation. A thousand years of deposition in such a system might amount to no more than a spadefull of windblown dirt. Even if it occurred "only" 10,000 years ago, which is relatively young for geological beds, it seems improbable that much sedimentological accumulation—and thus much of a fossil record of these events—would have been preserved. Once again we see a striking parallel with the dinosaur extinctions. "Where are the bodies?" cry the opponents of overkill.

The proponents of overkill believe that the extinction of larger land mammals in North America was rapid and devastating. According to Martin and others, 33 genera, spread out over a giant continent, disappeared forever during a 1000 to 2000-year interval. By comparison, only 20 genera of large mammals had gone extinct in the 3 *million* years prior to the great North American extinction.

Not all the larger mammals of North America went extinct, for 12 genera are still extant. But all these survivors have something significant in common: All were late arrivals to North America, reaching the new continent by the same land bridge between Siberia and Alaska that the Clovis traveled and coming, as they did, from either Europe or Asia. All of these mammals had long experience with humans. The deer, bears, puma, and bison now considered natives of North America were then every bit as foreign to this continent as the humans arriving in their midst. In reality, the extinction of large mammals in North America was essentially complete: None of the natives survived.

The great Scandinavian paleontologist Bjorn Kurten took note of this connection several decades ago. "It is noteworthy," he said, "that most of the Eurasian invaders of North America, the moose, wapiti, caribou, musk ox, grizzly bears, and so on, were able to maintain themselves, perhaps because of their long previous conditioning to man." Martin agrees. He views the survivors as more gracile and wary than those killed off. They are unpredictable in their movements and difficult to hunt. In his view, behemoths such as the ponderous mastodons and mammoths, gargantuan but slow ground sloths, and giant camels were easy targets for the nomadic Clovis people, themselves survivors of the harsh Ice Age. And as the giant herbivores disappeared, a suite of great carnivores also vanished, including a North America lion, the huge dire wolf, giant bears, and, perhaps most fearsome of all, the saber-toothed tigers.

. . . .

One of the critical questions related to the extinction of the North American large-mammal fauna is whether that fauna was in decline when the Clovis hunters first arrived. This question is also at the heart of the mystery surrounding the extinction of the dinosaurs. Was an abrupt catastrophy (the great comet at the end of the Cretaceous, the arrival of humans at the end of the Ice Age) simply the last straw for an already-stressed population that would have succumbed to mass extinction anyway? This question, so relevant to solving the mystery, is not very amenable to paleontological inquiry, for as Paul Martin points out, "Bones do not provide reliable estimates of past biomass." Nevertheless, Martin suggests that there is a real difference between the fates of larger animals in the Old and New Worlds. Part of the difference may be related to the differing histories of human and mammoth in these two great regions.

First and most important is the realization that mammoths, mastodons, and other large mammals died out earlier in Europe and much of Asia than in North America. The last European proboscideans disappeared 12,000 to 13,000 years ago, about the time that humans first arrived in North America, whereas in China they were gone as early as 20,000 years ago. Only in parts of Siberia and northernmost Asia did the large elephants last to even 10,000 years ago. It is also clear that humans had a long cultural history of sometimes utilizing elephants not only for food but also for materials used in shelter and tools.

Perhaps the most famous examples of this use come from the mammoth bone huts known from eastern Europe and northern Asia. This usage of bone material seems to be related to a striking paleontological phenomenon that is absent or very rare in North and South America. Huge numbers of mammoth bones are found in late-Pleistocene sites of central and eastern Europe and in many regions in Asia. Abundant human artifactual remains are associated with these sites, indicating that humans were frequent visitors to the great elephant graveyards. But were we simply visitors, scavengers, and gatherers, or were we responsible at least in part for the elephants' death? Human behavior is reflected in the assemblages. Most of these sites are probably not actual death sites of the elephants but human encampments or processing stations. They fire the imagination: great herds

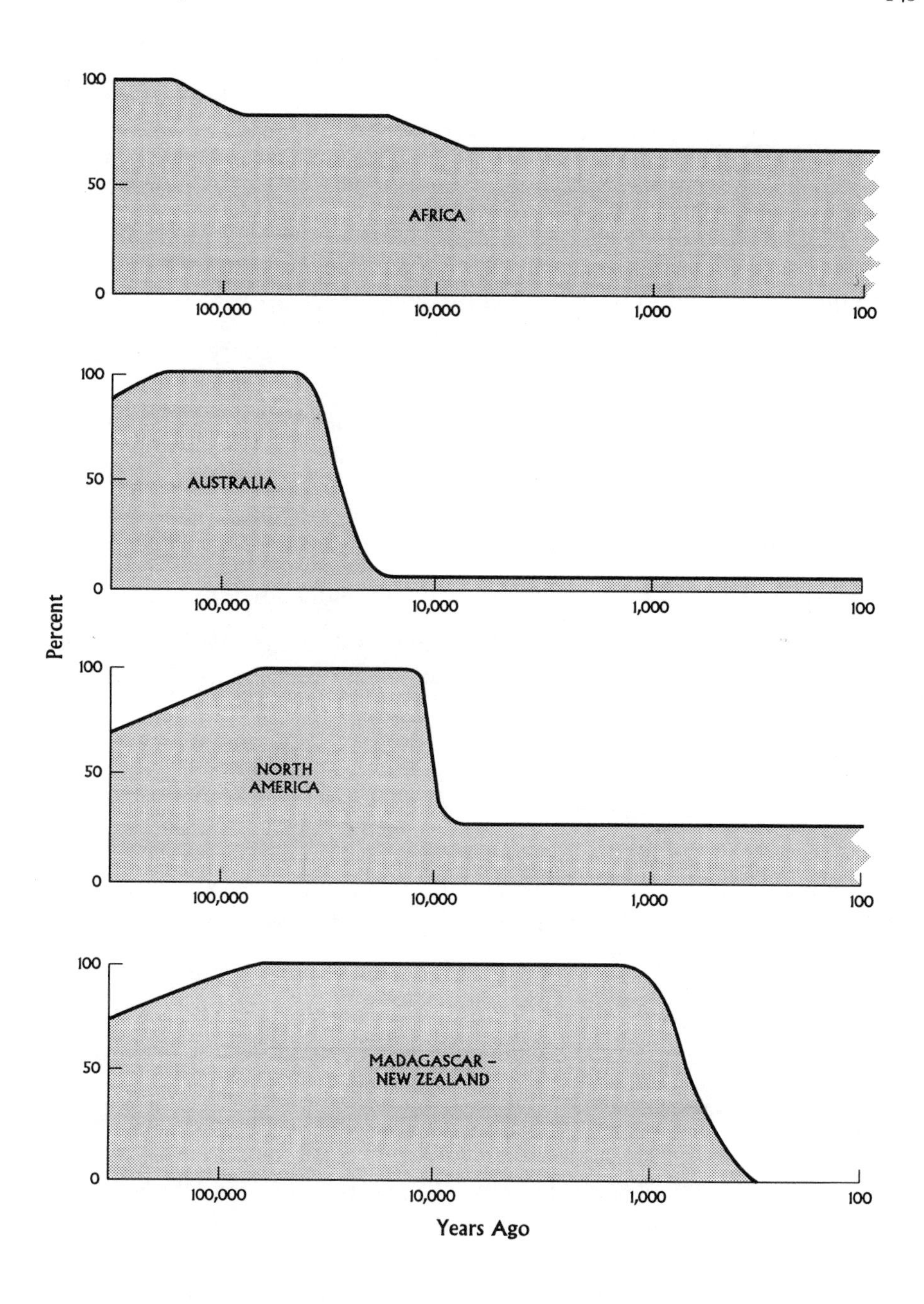

Disappearance of large animals in various continents. (Modified from Martin and Klein, eds., Quaternary Extinctions, 1989; with permission.)

of mammoths slaughtered, and their bones so thick after the hunt that they can even be used to build houses. Yet such a view may be hopelessly romantic and quite wrong.

More than 70 mammoth bone huts or other dwellings have been found at about 15 different sites on the central Russian plain. Although mammoth bone huts are known to have been constructed by Neanderthal peoples, the majority of these structures were built by us, *Homo sapiens sapiens*. They were built starting around 30,000 years ago, and the last is about 14,000 years old. Thus it dates from about the time when mammoths began to become scarce in Europe and central Asia.

The huts must have been extraordinary structures. Most were round and about 15 to 20 feet across at their base. The foundations were made of massive elephant skulls, jawbones, and shoulder blades, and the roofs and walls were made of lighter bones, tusks, and wood. Skins must have been stretched overhead to provide shelter. The use of these bones made sense, for the huts were constructed on plains and steppes barren of trees. A large fireplace was found inside, and bone as well as wood was burned for fuel.

These were not light houses. Archaeologists in Russia have counted and weighed the bones incorporated into each hut. At the site named Mezirish, a single hut was found to contain 385 mammoth bones with a combined weight of 20 tons. The entire cluster of huts at this locality contains bones from at least 150 individual mammoths. These huts were built for permanence. Some of the sites may have been used for centuries or even millennia.

Though dramatic, the mammoth huts are rare. According to archaeologist Gary Haynes, they account for only a small percentage of similarly aged sites discovered in the same region. Haynes believes that the huts are not from human meals:

> Our ideas about the late Pleistocene people of northern Eurasia are anchored to their houses and larders. But their houses used to be live animals, and it cannot be spontaneously assumed that Eurasian foragers in frozen steppes speared them all so that they could shelter behind skulls and scapu-

> lae. . . . Archeologically, these are huge, rich sites, with abundant artifacts, stacked tusks and bones from dozens of mammoths, with dates clustering at 15,000 to 25,000 BP. The archaeological consensus has traditionally been that these remains are from animals that had been hunted, butchered, and transported back to the hut sites. A new view is that the bones are about opportunistic building material. (Haynes, 1991)

Could any predator—other than us—have successfully killed an adult mammoth or mastodon? The young and the old would surely have been prey targets to the fierce large carnivores of the late Ice Age. But the adult elephants? It may be that when adult, these creatures were essentially invulnerable to all but human predation. And if so, the way in which human hunters interacted with the large elephants in the Old and New Worlds may have played a large role in their fate.

Bjorn Kurten was perhaps the first to suggest that the size and abundance of carnivores can serve as an interesting clue to the size and abundance of large animals. Kurten reasoned that a reduction in either of these characteristics in large prey would lead to their similar reduction in carnivores. Just such a change was observed by Kurten in the abundance and size of Ice Age carnivores in the Middle East, and Martin interprets this as the result of a long, slow extinction among these carnivores' larger prey—exactly the result that might be expected if long-term hunting were a factor in the diminishing number of larger prey. But *no* such changes have yet been discovered in any assemblage of carnivores from North America—just as there appears to be no scarcity of mastodon, mammoth, horse, or camel bones in any North America deposits. Thus, there is no evidence that the late-Pleistocene megafauna of North America were declining in numbers or diversity before 12,000 years ago.

Let us assume that the newly arrived Clovis people found a continent rich in big game. How long might it have taken them to populate such a large, empty continent? In a series of articles, Martin has made estimates of this. Central to any such estimate is the amount of land necessary to support one human. Our world of teeming cites and gigantic corporate farms supports a vastly larger population per square mile that was possible in the Pleistocene era, when society was entirely dependent on big game. Anthropologists estimate that even land rich in game can fill at best one human mouth for each square mile—and where game is scarce, far more territory is required. There are about a million square miles in continental North America south of the Pleistocene limits of the glaciers 12,000 years ago. North America could, according to these assumptions, support about a million humans if they required fresh meat as their primary food source.

According to this model, how rapidly would the continent have filled with humans? Let us assume that the original Clovis arriving from the north were only about 100 strong. If its growth rate was 0.1% annually, then the human population of North America would have reached the projected population of native Americans about the time when Europeans arrived, around A.D. 1500. Yet 0.1% represents a very slow population increase and seems to fly in the face of the history of humans arriving in lush regions. In such places as Hawaii and Pitcairn island, reliable records indicate that humans bred quickly and effectively. There were certainly dangers in the New World, and life expectancy could not have been long. But on the other hand, there were no plagues that we know of. Large-scale mortality from diseases seems to have originated in the Old World tropics and may have been largely unknown in the New World—at least until the Europeans arrived some 12 millennia later. Growth rates of the human population may have been much higher.

How fast can human populations grow? The highest rates today are observed in the Philippine Islands, where the human population doubles every 22 years. This is an annual growth rate of about 3.3%. If the Clovis showed such high growth rates, they would have saturated the continent to the theoretical carrying capacity of one human for each square mile in about 350 years—only about 20 generations. Even at a much lower rate of about 1.4% annually, or a doubling every 50 years, saturation would have required only 800 years.

Having only one human for every square mile does not seem like much of a population crunch. Only a million humans in North America, which today supports a population of at least 350 million that will soon approach half a billion, would surely seem like a very low population. No crowds on the beach, no rush hours anywhere on the continent! Yet, one million humans would surely have made a big dent in the large-mammal populations if all of those humans were dependent on big game for food.

The peopling of the continent took place in much the same way as the later "re-peopling" by Europeans. Bit by bit, a wave of population moved outward from the initial colonization point. The impetus for this movement is always the same: a search for new resources for a better life. In this we humans are not nearly so unique as we think. The outward migration of the Clovis, like that of any new group of immigrant humans, seems to have followed patterns exhibited by all groups of exotic animals unleashed for the first time in a virgin habitat. They moved outward in what can be called a *front*.

. . . .

Paul Martin has described in detail the way in which he believes a human front swept through North America—and left a wave of extinction behind it. Martin argues that the zone of high human population occurred only along the periphery of the zone of human occupation. Unlike the second North American population explosion, in which Europeans advanced from east to west but left behind cities and towns sustained by farmlands and centers of industry, the Clovis, with no ability to farm or gain sustenance other than through harvesting large game, may have left behind regions largely devoid of humans. There would always have been smaller game to hunt, of course, just as pockets of deer and smaller mammals remain today, even in regions densely populated with humans. But if Martin and other proponents of the overkill hypothesis are correct, these would not have been favored game of the Clovis, who were, after all, the not-too-distant descendants of Asians who subsisted on mammoths and mastodons. That tradition may have died hard, just as it did for the obligate buffalo-hunting Amerindian tribes of the 19th century.

The advance rate of the Clovis hunting front would have been determined both by the abundance of game and by the rate at which the nonhunting portions of the population—the aged, children, and mothers with very young children—could be moved. Martin argues that within a decade or less, populations of vulnerable large animals would have been either severely reduced or hunted to local extinction. As the quarry vanished through excessive hunting, the front moved on. In this way, the peopling of North America by the Clovis would have occurred as a slowly advancing wave: for the humans, a wave of ever-burgeoning human population; for the quarry, a wave of extinction.

. . . .

Throw a pebble into a small, still pond. The ripples from the splash spread outward, but as they move away from the source, they diminish in size. Now imagine instead a wave that increases in size as it moves outward.

Martin assumes that a band of 100 hunters arrived in what is now Edmonton, Canada, about 12,000 years ago. They encountered a land filled with beasts that had never seen humans and were thus little wary of these new bipeds. And the temperatures were far less extreme than in the present: cooler in summer and warmer in winter than today. Such a place must have seemed like paradise to a people whose entire mode of life depended on meat. The small group of humans would have thrived and multiplied.

In one of Martin's early models of the front, he made several assumptions about the rate of population growth and the speed at which the front advanced. Assuming an initial population of about 100 that arrived in Edmonton 11,500 years ago, a population growth rate such that the number of people doubled every 20 years, and a rate of advance of 10 miles each year, Martin estimated that the front took only about 350 years to reach the Gulf of Mexico. And by that time the population of North America had risen to a half-million or more, and most of the larger game in North America had been butchered.

This model is breathtaking in its simplicity and very difficult either to prove or to discredit. However, it is consistent with various lines of evidence, and its very plausibility—the fact that in more than 25 years no one has been able to disprove it—lends credence to this most unsettling of hypotheses. Unsettling, at least to me, for several reasons. If Martin is correct, then the earliest humans in North America underwent a rapid population explosion. One consequence was extreme predation on favored food items. The "normal" relationship between prey and predator, where long-term stability or at least fluctuating relative numbers prevail—did not apply.

Many of us harbor a romantic image of humans coexisting with nature in what we like to believe is the "native" state. Many models come to mind, but I think especially of the American Plains indians living on, and with, the large herds of bison. My image is of careful stewardship, where only enough animals are taken to ensure sufficient food and skins for clothing or shelter and there is neither "sport hunting" nor wasteful consumption. Yet such a view may be very naive. The Plains Indians riding horses in pursuit of game were very much a new phenomenon, for horses were not reintroduced to North America until 1500 and did not become common until the 18th or 19th century. How long would the great buffalo herds have lasted under the hunting pressure of mounted Indians?

There is another ominous aspect to the arrival of the Clovis: They might have suffered from what is called protein poisoning. This is a nutritional ailment found among Bushmen of the Kalahari, who live in an environment largely devoid of vegetation. An all-meat diet is not ideal for humans, and protein poisoning results when a regime of very lean meat is employed for long periods. The human body needs carbohydrates and fats, and lean meat has little of either. If the arriving Clovis were suffering from protein poisoning, they might have sought out large animals, such as the elephants, and eaten only those parts that were marbled with fat, such as the heart and perhaps the kidneys or liver. Bushmen suffering from protein poisoning do this, causing great wastage of prey, for many prey must be killed to extract enough fatty organs to feed a band of humans.

The front did not stop when it reached the Rio Grande or even Central America. According to Martin's model, humans reached Panama by 11,000 years ago; and by 10,500 years ago, our species reached the southernmost part of South America, wiping out larger mammals on that continent as well.

Martin summarized this sequence in chilling fashion:

> Unless one insists on believing that Paleolithic invaders lost enthusiasm for the hunt and rapidly became vegetarians by choice as they moved south from Beringia, or that they knew and practiced a sophisticated, sustained-yield harvest of their prey, one would have no difficulty predicting the swift extermination of the more conspicuous native American large animals. I do not discount the possibility of disruptive side effects, perhaps caused by the destruction of habitat by man-made fires. But a very large biomass, even the 2.3 billion metric tons of domestic animals now ranging the continent, could be overkilled within 1000 years by a human population never exceeding 1,000,000. We need only assume that a relatively innocent prey was suddenly exposed to a new and thoroughly superior predator, a hunter who preferred killing and persisted in killing animals as long as they were available.

At the end of it all, when the mammoths, mastodons, sloths, and other giant game were gone, one last group went extinct: the Clovis. Martin's model supposes that the population of humans precipitously dropped and that a great cultural evolution ensued. The Clovis culture was supplanted by many smaller groups, all faced with making a living on other food and exploiting new types of resources, such as the surviving mammals, the abundant fish, and a revolutionary concept: agriculture.

Martin not only described in words how he thought the front moved. He also tried to make mathematical models showing how it might have progressed. These and other models are profiled in the next chapter.

. . . .

The overkill hypothesis should apply to areas other than North America, and one test of the hypothesis is whether equivalent extinctions happened in other regions soon after the arrival of humanity. Paul Martin argues that this pattern occurred in South America, Australia, Madagascar, New Zealand, and Hawaii.

South America was separated from Central and North America by a deep expanse of the sea during the Cenozoic Era, so its fauna had a quite separate evolutionary history from the creatures of North America until the Isthmus of Panama formed some 2.5 million years ago. Many large and peculiar mammals evolved there, including enormous armadillo-like creatures called glyptodons and giant sloths (both of which later migrated northward and become common in North America), as well as giant pigs, llamas, huge rodents, and some strange marsupials. With the formation of the new land bridge in what is now Panama, just before the onset of the Ice Age, came a rapid faunal exchange.

As in North America, large-mammal extinction occurred in South America soon after the end of the Ice Age. Forty-six genera are now known to have gone extinct sometime in the last 15,000 years, and most or all of these appear to have been completed by 10,000 years ago—soon after the arrival of humankind. The results seem to be in accordance with Martin's predictions, and if anything, the extinctions in South America were even more devastating than those in North America.

Of all the continents, Australia suffered the greatest proportional loss of its megafauna. The tragedy of Australia's loss was the uniqueness of the extinct animals. The Australian continent, cut off from the mainstream of Cenozoic Era mammals, was the center of marsupial mammalian evolution, and it was among these extraordinary giant marsupials that rapid extinctions occurred.

The mass extinction that struck the Australian fauna during the last 50,000 years left only four species of large native mammals alive, and no new arrivals bolstered the disappearing Australian fauna. Thirteen genera of marsupial mammals, comprising as many as 45 species, disappeared from the continent. The victims included large koalas, several species of hippo-sized herbivores called *Diprotodon*, several giant kangaroos, several giant wombats, and a group of deer-like marsupials. Marsupial carnivores were lost as well, including a large lion-like creature and a dog-like carnivore. In more recent times a third predator, a cat-like creature found on offshore islands, also vanished. Other victims included a giant monitor lizard, a giant land tortoise, a giant snake, and several species of large flightless birds. The larger creatures that did survive were those that were capable of speed or had nocturnal habits.

Paul Martin believes it is no coincidence that just as humanity reached

Australia much earlier than it reached North or South America, so too did the wave of extinctions that assailed the Australian megafauna begin earlier than in the Americas. Reliable evidence now shows that humankind reached Australia no later than about 35,000 years ago, and some archaeologists think we were present in Australia as early as 50,000 years ago. Most of the larger Australian mammals were extinct by about 30,000 to 20,000 years ago. And in Australia there were no glaciers and no sudden warming event such as occurred in North America and Europe. The climate changes that took place were nowhere near as dramatic as in North America, and yet the rate of extinction was much higher.

A different pattern emerges in the areas where humankind has had a long history, such as Africa, Asia, and Europe. Martin predicted that in regions humanity has long inhabited, fewer extinctions than in the Americas or Australia should have occurred, because the hunters and hunted would have shared many tens of thousands of years of co-evolution. This is in fact what we see. In Africa, extinctions occurred 2.5 million years ago, but later losses were far less severe than those in other regions. Africa did not go unscathed, however; the mammals of northern Africa, in particular, were devastated by the changes in climate that gave rise to the Sahara. In eastern Africa, little extinction occurred, but in southern Africa, significant climate changes 12,000 to 9,000 years ago coincided with the extinction of six species of large mammals. Europe and Asia also suffered fewer extinctions than the Americas and Australia; the major victims were the giant mammoths, mastodons, and woolly rhinos.

Paul Martin and other adherents to the overkill hypothesis have amassed a tremendous amount of information and data in support of their theory. Martin considers eight attributes of the Ice Age extinctions to be especially important.

1. *Large mammals were the primary creatures going extinct.* This is perhaps the most salient aspect of their extinction. It was not the small animals but the larger ones. And it was not just mammals: On both Madagascar and New Zealand, where the largest animals were birds, it was birds—not

mammals, that went extinct. In this, the Ice Age extinction closely matched the great extinction at the end of the Mesozoic. Mammal species with average weight of 100 pounds or more showed the highest rates of extinction, just as during the Mesozoic it was the largest land animals, the dinosaurs, that showed the highest extinction rate.

2. *There were very few extinctions among small mammals while the larger species were disappearing.* This point may be not so well known. Although we have a good record among smaller mammals, few bird species are known from the Ice Age, because bird bones fossilize more rarely than large mammal bones. Smaller animals apart from the mammals are even more poorly known: There might have been enormous extinction among insects, snakes, frogs, and other smaller creatures, and we would have only scant information about it. Nevertheless, it does looks as though larger animals had a much higher extinction rate.

3. *Large mammals survived best in Africa.* The loss of large mammalian genera in the last 100,000 years was 73% in North America, 79% in South America, and 86% in Australia, but in Africa only 14% died out.

4. *Extinctions could be sudden.* One of the most surprising and disturbing features of the Ice Age extinctions was how rapidly entire species could be lost. Much of the debate over extinctions in the deep past is related to extinction rates over time. We simply do not have the technology to discriminate even thousand- to ten-thousand-year blocks of time in such ancient rocks. For the Ice Age, however, carbon dating techniques allow very high time resolution. These techniques have shown that some species of large mammals may have gone completely extinct in periods of less than 300 years!

5. *The extinctions occurred at different times in different places.* Unlike the dinosaur-killing Cretaceous extinction, in which the final die-offs took place simultaneously all over the earth, the Ice Age extinctions took place at different times. In the Americas, they occurred about 11,000 years ago, in Australia perhaps 30,000 years ago.

6. *The extinctions were not the result of invasions by new groups of animals (other than humankind).* It has long been thought that many extinctions

take place when more highly evolved or better-adapted creatures suddenly arrive in new environments. This was not the case in the Ice Age extinctions, for nowhere can the arrival of some new nonhuman fauna be linked to extinctions among forms already living in the given region.

7. *Extinctions occurred soon after the arrival of humankind.* There was very little time lag between humans' arrival and the ensuing extinctions.

8. *The archaeology of the extinctions is obscure.* Few archaeological sites in North and South America yield remains of extinct creatures. Only mammoths and mastodons have been found in these kill sites, and in Australia no kill sites at all have been found. In temperate parts of Eurasia, on the other hand, Paleolithic artifacts are commonly associated with bones of large animals. In Eurasia, only four genera of large mammals were lost late in the Ice Age: the mammoth, woolly rhinoceros, giant deer, and musk ox. Critics of the overkill hypothesis have often pointed to this aspect as the most powerful argument against overkill. Martin and others, however, view the Eurasian findings in a different light. They believe extinctions happened so quickly that there is only a small window of time containing sites with evidence that "humans did it." In 1973, Martin noted, "The scarcity of kill sites on a landmass which suffered major megafaunal losses becomes a predictable condition of the special circumstances which distinguish a sudden invasion from more gradual prehistoric cultural changes *in situ*. Perhaps the only remarkable aspect of New World archeology is that *any* kill sites have been found."

Furthermore, given the amount of time necessary to produce a good blade or weapon capable of killing a large mammal, why would the ancient hunters leave these valuable tools with their dead prey? Perhaps we should wonder why we have found so many spearpoints associated with prey, rather than why we have found so few. A spearpoint is not like a bullet. Unless it is broken, it can be reused instantly, and even broken blades can be fashioned into spearpoints again through renewed chipping. Especially as stone appropriate for tips became scarce in North America, which appears to have happened after 10,000 years ago, the spearpoints used in the hunts would have been prized commodities.

The overkill hypothesis has several testable attributes: Extinctions should occur soon after humanity's arrival in a new, uncolonized continent; endemic species of large size (human food sources) that have no experience with humanity should be the most endangered; the extinction should take place in a relatively short time, and should proceed as a wave sweeping across the continent.

. . . .

It is a rare theory that does not meet with immediate scientific opposition, and a theory as old and contested as the overkill hypothesis was fair game. Overkill is one of two leading hypotheses concerning the fate of Ice Age megafauna. The other is that the climate change that occurred at the end of the Ice Age was the assassin.

There is a long history of suggestions that slow climate change caused the faunal turnover at the end of the Ice Age. In many ways this explanation seemed to fit the evidence better, and it may have been psychologically far more palatable. Here is a scenario about how climate change may have acted as a big-game killer, at least in North and South America. As the glaciers receded, a profound reorganization occurred in hydrological, biological, and sedimentological systems. The world's ice coverage changed from 30% of its surface to the current level of about 10%. This phase change of so much water from solid to liquid was the driving force behind much of the subsequent geological, climatic, and biotic transformation. As the ice melted, water and sediment choked the rivers; they altered the drainage of great lakes and, in breaking the ice dams that had formed some of these lakes, caused cataclysmic floods that themselves changed the face of North America (the channeled scab lands of Washington State and Idaho are examples). The continental shelves changed positions and depth as the shorelines migrated inward under rapidly rising seas. Rainfall patterns shifted; the seasons themselves were lengthened or shortened. Temperatures rose all over the earth. And of course, in the wake of so much physical perturbation to the environment, biotic systems changed.

These great changes in the physical environment are at the heart of many scientists' belief that the large-animal extinction at the end of the Ice Age was brought about by physical or environmental changes, not by human hunting. The extinction of many species in a broad environment must involve many variables in both biotic and physical spheres, because ecosystems are complex. Physical changes may bring about the death of one species, and its disappearance then affects other species, such as its predators, prey, and parasites. In this way even a single extinction causes effects that ripple through the ecosystem. At the end of the age of glaciation, many species faced environmental changes, and surely many succumbed before they could adapt or migrate. Habitat destruction is probably the major cause of physically induced extinction.

The extinction of so many large Ice Age mammals very much resembles the extinction of the dinosaurs. All were big, all were very important members of their communities (in the sense that their passing readily transformed the communities), and all died out under mysterious circumstances. There is another similarity: The leading explanation for both the death of the dinosaurs and the death of the Ice Age mammals has been slow climate change. Not that there has been much direct evidence that climate actually did it, but it has long been believed that the duration of the extinction event itself was protracted and thus must have been related to some gradual environmental change.

The most detailed of the climate-induced models of extinction does not rely on sudden temperature or moisture change *per se*. Developed by R. Graham and E. Lundelius, this model suggests that the cool but stable conditions known to have characterized the late glacial period changed to warmer but more extreme temperature regimes following the glacial retreats. These climate changes caused North America's various plant communities to became less diverse and thus less able to support a diverse assemblage of mammals. Small mammals migrated to new regions, but large mammals, requiring more food, died out. There is no doubt that the end of the Ice Age was accompanied by sudden, drastic changes in temperature and that a dramatic change in plant communities and their distributions across North America soon followed. But the idea that none of the larger mammals was able to migrate out of harm's way seems implausible; we know that many large African mammals are perfectly capable of long treks in search of seasonal food or water. Climate change alone seems unlikely to have killed off 35 genera of North American mammals in just 2 to 5 millennia.

There is some evidence that the amount of seasonal change affecting terrestrial communities, at least in North America, did differ from previous interglacial intervals. But apparently not by much. The evidence comes from the distribution of various animal types, such as large turtles, that inhabited large areas of North America in previous interglacial periods but no longer do so. The inference is that during previous interglacial intervals, climates were warmer than they are today. But in many scientist's minds this difference is not enough to account for the death of so many larger animals. Another distinct or complementary mechanism must have been involved.

One argument concerning the fate of the Pleistocene faunas has been proposed by Russell Graham, a scientist at Illinois State University. Graham attributes their demise neither to the spears of hunters nor to climate change. He invokes a far more insidious cause: changes in food supplies as the great steppes, once dominated by a complex and diverse assemblage of herbs and shrubs, were transformed into immense and far less nutritious grasslands.

This seems like heresy: To humans of the industrial age, wide fields of grasses connote richness and productivity. We have seen so many films of great buffalo herds grazing contentedly on the vast plains of the American midcontinent. The reality is that these grasslands (at least before farmers transformed them through selective breeding and the introduction of new species) may have been a far poorer food source than what they replaced. Vegetation of the Ice Age appears to have been more diverse and more nutritious for a large variety of herbivorous mammals than either the grasslands or the forest that replaced them.

The nutritional value of native vegetation depends on several factors, including climate (which determines the length of the growing season) and the nutritional availability and recycling of the soil. The makeup of the plant species within communities is also important. Plants utilize several different photosynthetic pathways, and a plant's nutritional value to herbivorous mammals is related to the pathways it uses. Grasses, for instance, can be characterized as having either a three-carbon or a four-carbon cycle of photosynthesis, and C3 grasses yield more nutrients for large mammalian herbivores. The relative distribution of these plant types seems to be strongly influenced by climatic factors. C3 grasses generally inhabit cooler, moist environments whereas C4 grasses favor warmer, drier environments. As long ago as 1980, it was hypothesized that the end of the Ice Age was accompanied by a climate transition favoring C4 grasses as the northern hemisphere became warmer and drier. The transition from grasslands dominated by the more nutritious C3 grasses to those with the less favorable C4 types would have lessened the carrying capacity of large segments of Earth's surface, reducing the overall biomass of herbivores. Because the climate changes appear to have progressed from south to north, this transition would have forced the communities of mammals to move northward in search of food.

Changes in plant assemblages may have brought about more than a lessening of nutritional value. The end of the Ice Age may have favored the spread of new types of plants actually toxic to larger herbivores. The grasses found today in the far northern tundra regions, where gigantic herds of megamammals once lived, contain a wide variety of mildly toxic alkaloid substances. Alkaloids are the chemical substances that make eucalyptus trees so fragrant; in higher concentrations, however, these compounds can be poisonous. Their presence would thus limit the nutritional value of grasses containing them.

Browsing herbivores also may have experienced a decrease in nutrient availability at the end of the Pleistocene. Russell Graham points out that browsers—herbivores that eat the tender new growth of trees and bushes rather than grazing exclusively on grasses—also encountered a gradient in nutritional value depending on the plant species eaten. Trees and shrubs generally use two types of defenses against browsing. Some species simply regenerate lost material through rapid growth; others employ a more active defense by secreting toxic resins that, like the alkaloids in certain grass species, reduce the palatability of the plant. Graham argues that vegetation changes accompanying the climate changes at the end of the Ice Age increased the number of low-nutrition trees and shrubs available to the larger

Pleistocene herbivores. Over vast regions of the northern hemisphere, plant assemblages dominated by highly nutritional willow, aspen, and birch trees changed to far less nutritious spruce and alder groves. Of course, even in areas dominated by spruce, a relatively non-nutritious tree, many nutritious plants were still available. However, as climate changes reduced the number of nutritious plants, the herbivorous mammals would have increasingly foraged on the remaining high-nutrition plant types, thus hastening their demise—and perhaps leading to the reduction in size of many mammal species dependent on vegetation.

As the Pleistocene ended, the more open, high-diversity spruce forests and nutritional grass assemblages were rapidly replaced by denser forests of lower diversity and lower nutritional value. In the eastern parts of North America the spruce stands changed to large, slow-growing hardwoods such as oak, hickory, and pine, while in the northwest the great forests of Douglas fir covered the landscape. These types of forests have far lower carrying capacities for large mammals than the Pleistocene vegetation that preceded them.

By the early 1990s, various models in which climate change either directly or indirectly caused large-animal extinction posed a formidable alternative to the overkill hypothesis.

. . . .

Just as Paul Martin has become the most visible spokesman for the overkill theory, so too has a single individual come to represent those scientists who believe that climate change was the primary cause of the Ice Age extinctions: Don Grayson, an archaeologist at the University of Washington. Grayson has worked for many years on the paleontology of the southwest, and like Martin, he has an intimate knowledge of paleontology, archaeology, and geochronology, the methodology behind carbon-14 dating.

Grayson's major critique of the overkill hypothesis focuses on what he sees as problems in the chronology of the extinctions, and especially in the validity of carbon-14 measurements made decades ago. The overkill hypothesis depends on two assumptions: that most or all of the large animals died out in a short, 2,000-year period between 12,000 and 10,000 years ago, and that no extinctions took place *after* 10,000 years ago. Grayson believes both of these assumptions may be false.

The data set recording the chronology of Ice Age mammal extinction comes from a great number of radiocarbon dates accumulated between the late 1950s and the present. However, the methodology of carbon-14 dating vastly improved during that interval, making later age determinations far more accurate. Grayson believes that many dates arrived at in the 1960s and 1970s contain errors, perhaps quite large ones, due to imprecise laboratory techniques. He has culled earlier radiocarbon dates from the entire data set and come to the conclusion that the bulk of the extinctions of large Ice Age mammals did *not* take place during the critical interval between 12,000 and 10,000 years ago. Rather, he maintains, many species were already extinct by the time the first people arrived in North America some 12,000 years ago. The same argument can be applied to South America. Grayson also voiced another common criticism of the overkill scenario, an argument analogous to that of the dinosaur workers who seek Cretaceous "bone beds": Very few mammal skeletons have actually been found in "kill sites," places where the fossils of extinct species are associated either with human artifacts used to kill the animal or with such evidence of human activity as bone butchering. Kill sites remain rare, and very few extinct mammal species can be found at such sites. The implication is that if early people were slaughtering animals in such numbers that rapid extinction ensued, then either they were doing it very covertly or it happened during a very narrow window of time.

Grayson and others in his camp thus argue that the extinctions were the result of the profound—but gradual—climate changes that accompanied the retreat of the glaciers. There is no doubt that great changes in climate were taking place around the globe. But did these climate changes kill anything? The end of the last great glaciation was an Earth-changing event, yet it was not unique. The ice ages have been under way for much of the last 2 million years, during which time there have been numerous episodes of glacial ebb and flow. Yet there have *not* been great animal extinctions each time. In fact, the last million or so years have seen at least 22 alternating stages of high and low ice volume in North America but only two significant extinctions, the one of 10,000 years ago being the largest. Was this last major episode of climate change somehow different from all the others? Until the early 1990s, no such difference could be discerned, and this absence of clear change in climate greatly detracted from the arguments of Grayson and his supporters.

Grayson's last major contribution to the debate was a paper published in 1991, in which he argues that the bulk of extinctions of the larger mammals of North America (35 genera) occurred prior to 12,000 years ago, well before the incursion of the Clovis hunters into North America. His argument was based on new interpretations of the carbon-14 dates used to provide a chronology for the extinctions. Grayson pointed out that only the more recently completed analyses (from the late 1970s onward) could be considered precise enough to be used. He concluded this article with a summary of his position:

> The results of that search [for reliable radiometric age dates on the extinct mammals] strongly suggest that overkill could not have been the force that Martin has claimed. The differential appearance of kill sites (only proboscideans, and within the proboscideans, almost only mammoth) and the strong hints that many of the taxa involved may have been on their way to extinction, if not already gone, by 12,000 years ago imply a far lesser human role in the extinction than the overkill model allows. The climatic models account not only for the extinctions, but for the histories of smaller mammals during the Pleistocene. With greater explanatory power, most scientists studying the extinctions issue accept climatic, not overkill, accounts, while recognizing that far more precision is needed in these accounts. This does not mean that people played no role in causing the extinctions. A multivariate explanation may yet provide the best account of the extinctions. But no matter what the human role might have been, overkill was not the prime cause of the extinctions. That cause rather clearly lies in the massive climate change that marks the end of the Pleistocene in North America. (Grayson, 1991, p. 220)

Upon reading this in the mid-1990s, I was struck by the almost eerie similarity between the Grayson's arguments and those being leveled against the impact theory for the disappearance of the dinosaurs. The proponents of both arguments believe that the victims—the Pleistocene megamammals and the Late Cretaceous dinosaurs—were dwindling in diversity and abundance well before their extinction. Both assume that there would be a "bone bed", or that there would be more kill sites, if the sudden, catastrophic explanation were correct. Both cite last occurrence dates (the time when the last known individual of any species occurs in the fossil record) for the victims as being well before their supposed final extinction. And both imply that they are correct in their assumptions because "most scientist think they are correct," which suggests that establishing truth in science is a democratic process.

. . . .

The megamammal extinction is most commonly thought to have been caused by either overkill or climate change. It has long been assumed that one or the other of these hypotheses will eventually be proved false. However, no one seems to have considered the possibility that, at least as they were defined through the late 1980s, both hypotheses might be wrong.

In 1996 a new voice entered the debate. Paleobiologist Michael Beck published a paper in the journal *Paleobiology* directly confronting the overkill and climate change hypotheses. Beck described the problem as follows:

> Many hypotheses attempt to explain the brevity and severity of the megafaunal extinctions in North America at the end of the Pleistocene. Two primary causes form the poles of the debate: climate change and human predation. Climatic models suggest that large mammals, which require vast ranges of primarily food resources and have low intrinsic rates of increase, were unable to respond to rapid environmental changes during the deglaciation. Human predation models postulate that large prey including mastodons, mammoths, and camels were hunted to extinction by human migrants from Siberia.

Beck went on the to list the testable assumptions of the two models.

1. The human predation model, or the blitzkrieg hypothesis of Paul Martin and others, assumes that the megafaunal extinctions occurred abruptly at the end of the Pleistocene. Indeed, it was the apparent rapidity of the extinctions that convinced Martin and his colleagues that only human preda-

tion could account for such a rapid mass extinction. If, however, the extinctions could be shown to have occurred over a *long* period of time, then Martin's contention would be notably weaker. Don Grayson and others have raised just this objection, saying their culling of the radiocarbon data suggests that the extinctions did take place over a much longer time interval.

2. The blitzkrieg model proposes that the mass extinction occurred along an expanding "front," a wave of extinction beginning in Edmonton some 12,000 years ago and ending at the tip of South America about 2000 years later. This model thus predicts that the extinction would be "time transgressive" from north to south. That is, kill sites of large animals should be progressively younger as one moves southward in the two continents during this time interval.

Beck analyzed these two problems using data of previous workers, including carbon-14 data from the papers of both Paul Martin and Don Grayson. He came to some surprising conclusions. First, Beck analyzed for the duration of the extinctions: Were they rapid (occurring in less than 2000 years) for the various animals examined (11 genera of large megamammals)? Or were they spread out over several thousand years? Using sophisticated statistical tests, Beck concluded *that the extinctions took place rapidly*. He thus refuted Don Grayson's assertion that the extinctions took place over a longer period of time. So far so good for the overkill camp. But then Beck examined the assumption of the front, which predicted that kill sites farther from Edmonton would be younger. To the chagrin of the overkill camp, he found no such pattern! Two knockouts, and nobody left in the ring! In Beck's words, "I conclude that the reliable data [carbon-14 data] support a terminal megafaunal extinction." But then he goes on to say, "No matter how the data are analyzed, the geographic distribution of the terminal Wisconsin sites does not support the Blitzkrieg hypothesis. In fact, all of the patterns in the data are in opposition to the patterns predicted by the Blitzkrieg hypothesis and there are no obvious biases in the data that could create these results."

Beck's analysis strongly suggests that the concept of a front, an expanding wave of extinction, is incorrect. But it also shows that in one important respect the adherents of overkill are right: The extinctions, at least in North America, were not drawn out over many thousands of years but were very rapid—almost instantaneous—in geological terms, for a 1000 to 2000-year interval is virtually invisible in the stratigraphic record. Could it be that the real culprit was climate change, but climate change occurring so quickly that it produced a sudden, catastrophic extinction?

Of all the implications of the entire debate about the Ice Age extinction, this possibility is the most sobering. Imagine a climate change so severe that 70% of North America mammals were extinct in 1000 to 2000 years. Imagine there occurring, at the time of Christ's birth, some climate change so severe that by the present day, virtually every large mammal is extinct. What could such a change be, and if it occurred today, what would it do to the highly regulated and intricate human mechanisms of survival and to its agriculture? Any change that could create such a widespread dying could do so only by catastrophically perturbing large herbivore food supplies. Any such change would surely wreak havoc with our crops, their growth, and the planting schedule on which all of us depend.

Are there models of climate change that can even explain the terminal Ice Age extinction? Beck argues that there are not: "Grayson [in 1984] suggested that climatic models with strong predictions [for extinction] were being developed. More than a decade later, however, the paucity of well-defined, *a priori* predictions and true tests of these models suggests otherwise."

. . . .

Grayson's 1991 paper was his final salvo in a polite but earnest war with the overkill adherents, a scientific debate that had been under way for two decades or more. Paul Martin had delivered his final word even earlier, in 1986. Like many protracted engagements, it seemed to end in a stalemate. Dale Guthrie, in a paper published in 1990, said the debaters had "failed to reach a consensus on the causal linkages." Also like many such engagements, this one wound down because both sides had run out of things to say. Further hostilities required new weapons—new data or new ways to analyze old data. By the mid-1990s, just such new information and new methodologies became available. Some derived from new manipulation of existing data, and some came from unexpected sources: the dinosaur extinction debate, the ancient icepack covering Greenland, and conservation biology. But the most formidable advance came from a new mathematical modeling of population size, based on what we might call *kill curves*.

Mammoths killed by bears during the Ice Age. (From L. Figuier, *La Terre Avant Le Déluge*, 1864.)

7

The Kill Curve

New Orleans in early November: a mix of warm sun followed by more biting days, weather on the cusp between humid summer and cooler winter, between shirtsleeves and coats. My hotel was in the busy downtown core, a ten-block stroll from the French Quarter, and the walk was like the changing and uncertain weather. From a safe, homogeneous American chain hotel I walked out into the war zone, acceptable by day but spooky in the nocturnal hours when the discount shoe shops, secondhand clothing stores, and donut emporia are closed, and the crowds have changed from casually clad daytime workers to shabbily clad inner-city drug addicts and alcoholics, the predators who lean against buildings or fill the bus stops, forever waiting for some *other* bus. On my long solo walks through this district, watching these hominid birds of prey watch *me* as I pass, I think I can read their minds: *Is he a cop? Does he have money?* Or perhaps they are simply wondering how to get the next hit. New Orleans, a beautiful city that sometimes has the highest murder rate in America, during this late fall of 1995 was also temporary home to a swarm of geologists. More than 7000 had descended on the Big Easy for the annual meeting of the Geological Society of America, for four days of science, networking, drinking, and politicking. I came too.

I walked that long stretch, avoiding sales pitches ("It's all good!") and taxis. I was sad and angry, for one of my best friends had died only a few days before, overdosing on crack. He had had a medical degree and Ph.D. to boot, a beautiful fiancée, a house, a life—and an addiction. I had come to New Orleans to deliver a scientific address about a new way to model extinction, but my thoughts were about a simpler, more personal kind of death. These thoughts filled me with more rage than sorrow and I found myself plotting the extinction of the coca plant. There was no way to bring my friend back, and I wasn't even able to blame him yet. Though it was, after all, his addiction and his responsibility to seek help, I had only thoughts of revenge. And my fantasies of revenge focused not on the drug dealers or the South American cocaine cartel or even the endless chain of crooked politicians in so many countries, but on the plant itself. I hated this floral species that by sheer chance had evolved a chemical compound that, when ingested by humans, promises nirvana and brings only destruction. Cocaine, another random evolutionary step that brings biological success to one species and extraordinary misery to another. Thus my thoughts were on murder, but murder of a species, not an individual. Such murder has a name; we call it extinction.

Extinction is much harder to comprehend than simple murder. In my capacity as grieving friend, I wanted every single individual of the traitorous species to be eradicated at once; there could be no survivors. I tried to imagine finding every single coca tree throughout tropical South America and destroying every seed and every viable seedling. How many individuals are there in this plant species? Millions? Tens of millions? Yet the scientist in me argued that such wholesale extermination would not necessarily be required. Extinction is indeed mass murder, true genocide, or (perhaps more literally) eradication of a gene pool, but a series of new mathematical studies have shown that not every individual must be killed off to eradicate a species. Over the last several years, brilliant mathematicians have shown the existence of what is called a *minimum viable population*, a number below which no species can ever recover. It gave me some slight solace "*I don't have to kill off all of them, just most of them.*" Random forces would finish the rest, just as they did the mastodons and mammoths, the dinosaurs and ammonites.

With such morbid thoughts I walked the streets; sometimes the bolder vampires would stalk me, and I welcomed the excuse to turn and snarl at them. Such sad, hopeless thoughts we produce when our mourning cannot be consummated, true rage against the machine. Anger, not even an emotion, only the human mask for fear and hurt.

I was not the only angry paleontologist dwelling on extinction in New Orleans. I had come to deliver a talk about the dinosaur-killing K/T extinction in a symposium sponsored by the chief opponents of the meteor impact theory. Just as many dinosaur specialists refuse to believe their beloved saurians could have been swiftly executed by the Chicxulub comet, there are a few invertebrate paleontologists who still believe that the end-Mesozoic mass extinction was not a sudden catastrophe. Chief among this loyal opposition are Gerta Keller, a micropaleontologist from Princeton University, and Norman MacLeod, an American now working at the British Natural History Museum in London. Both had been actively investigating this extinction since the early 1980s, when they were among the vocal majority arguing against either an impact or an impact-generated extinction. Now they were the vocal minority, arguing even louder in concerted opposition to the impact hypothesis. They were not amused at being leaders of a cause with fewer adherents each year. The stakes, after all, are high. Who wants to go down in scientific history as backing the wrong horse in one of the most celebrated scientific debates of recent times? For Keller and

MacLeod it was round 11 of a scheduled 12-round fight, and they were now so far behind in points that only a knockout could save them. Increasingly of late, they had come out swinging.

Scientific talks in this venue are quite formal. Each scientist is given 15 minutes to make his or her presentation in a cavernous room that may hold up to 500 people. Yet while the speaker presents information, illustrated liberally by slides, people come and go, talk or take notes, nod off, or watch out for friends and colleagues.

My talk addressed a new theoretical methodology for studying and testing the nature of mass extinction. The only way to determine whether a mass extinction was sudden (and thus catastrophic) or more drawn out (and thus probably due to some other cause than one that would create a sudden extinction event) is to examine the rate at which species disappear. Yet the fossil record is notoriously uneven in its accuracy. Darwin knew this, and it a caused him much consternation. Darwin railed about the inadequacy of the fossil record, citing the paucity of fossils and their often-poor preservation. But perhaps most frustrating to him as to scientists these many decades later, was uncertainly about the true stratigraphic ranges of fossils. Where can they actually be found in sedimentary rocks? What is the chance that any fossil you find marks either the first or the last occurrence of that species? Very slim.

This argument remains central in the debate about dinosaur extinction, for the supposed layer containing the last dinosaur fossil is found as much as 2 to 3 meters below the comet-produced iridium horizon in Montana, the place where the last dinosaurs are best known. If the dinosaurs were killed off by the comet's collision with Earth, shouldn't their skeletons occur right up to this boundary and then disappear? In the early 1980s, this argument was repeated endlessly: Dinosaurs could not have been killed off by the effect of the comet because they were already extinct by the time the comet left its celestial scar on the planet. Other scientists, however, examined this premise, and when statisticians began to run the numbers on this question, they arrived at a seemingly illogical conclusion. They found that it is highly improbable that fossils will *ever* yield a true picture of extinction, even where the extinction is sudden. Various tests showed that even in cases where an extinction was known to have been abrupt for many species, imperfections of the fossil record made it look gradual. This concept, that a sudden extinction will always look gradual—that it will leave a record that suggests the fossils disappeared before the actual, sudden extinction horizon—is now called the Signor–Lipps effect after Professors Phil Signor and Jere Lipps of the University of California. Their fundamental discovery seems almost counterintuitive, yet it has transformed the way we analyze the data from mass extinctions.

It was just this argument—that the extinctions of the large Ice Age mammals looked gradual, not sudden—that had led many anthropologists to conclude that the cause of Ice Age mammal extinctions could not possibly be related to a sudden catastrophe, such as overkill by humans. However, the statistical work profiling the dinosaurs' demise suggests that a simple literal reading of the fossil record can lead to grievous errors of interpretation. The same may also be true for the Ice Age extinctions.

By the early 1990s, investigators realized that other statistical protocols might also yield insights into the tempo and mode of mass extinction, just as the Signor–Lipps effect had done. For example, can we assign statistical confidence intervals to determine the probability that the youngest or oldest fossil of a given species that we have found to date is *really* the youngest or oldest? In my talk, I re-examined the contention that the ammonite mollusks died out prior to the Chicxulub comet's impact. A brilliant mathematician from UCLA, Charles Marshall, had devised a set of new statistical protocols to test these assumptions. He created a series of confidence intervals on stratigraphic ranges that enabled scientists to test the probability that any fossil was indeed the last of its kind existing in the rock record. Charles and I had traveled to Spain and France the previous year to examine the ranges of ammonites there and thus test his methodology on rocks and fossils, and the meeting at New Orleans marked the unveiling of our findings. The work provided a way of actually testing hypotheses about extinctions.

My talk at New Orleans was to be the last delivered in a symposium assembled by Gerta Keller and Norm MacLeod and dominated speakers who were not exactly sympathetic to any concept of sudden extinction. Sitting in that large hall was like going back in time—back to the 1960s and the careful, conservative approach of geologists still committed to Uniformitarianism. Speaker after speaker put forth arguments against sudden extinction that had been refuted a decade ago. The two most common were that most species were found to occur last in sedimentary sections prior to the comet's impact and that most populations were dwindling in number before the impact anyway. One talk also asserted the presence of a bone bed near the K/T interval, although a critical examination of the data later showed this supposed bone bed to be nothing of the sort.

At the end of the session, I was finally to give my talk. Marshall's and my analyses indicated a strong probability that the K/T extinction was sudden and brutal, a short-lived event whose actual ferocity has been softened and clouded by the incomplete sampling of fossils prior to the iridium layer. These results directly contradicted the viewpoints of Keller and MacLeod.

. . . .

The dinosaur extinction debate has spawned many innovative approaches to paleontology and continues to do so. One of the most interesting we owe to David Raup of the University of Chicago. Raup proposed the concept of the *kill curve* in 1990. Trying to establish a mathematical relationship between the impact of a celestial body and the number of resulting species extinctions, Raup reasoned that the larger the impact, the greater the biological consequences. In a later treatment of this concept, in his excellent 1991 book *Extinction: Bad Genes or Bad Luck?*, Raup expanded on this concept to include not only asteroid and comet strikes but also other kinds of environmental degradation, such as temperature change and rising or falling sea level. From these analyses a powerful new way of looking at extinction was born. Kill curves can be calculated for a variety of causes, at a variety of taxonomic levels, and for a variety of time scales, but they all have one thing in common: They describe the death of organisms over time as a function of environmental change.

Such analyses are not important only for understanding past mass extinctions. One of the most controversial subjects in both paleontology and conservation biology today is whether or not our biosphere has entered, or is about to enter, a mass extinction. To answer this, we must construct a kill curve for the last 20,000 years and ask whether the rate of species death has been increasing, decreasing, or remaining steady. Yet gathering robust data to construct such a curve is more difficult than it might seem at first glance.

To construct a kill curve, we must first ascertain what the "normal" rate of extinction is. The fate of every species is to go extinct sooner or later, just as the fate of every individual is to die. Yet such is the longevity of most species (usually on the order of millions to tens of millions of years) that any given year should see only a handful of extinctions. During mass extinctions this rate increases drastically. The normal rate of extinction is called the *background extinction rate*, and the problem is to ascertain whether extinction rates that exceed this background level are chance events or indications of a trend.

Background extinction rate can be calculated by observing the number of extinctions occurring in any given million-year interval and then computing an extinction rate per year. The resulting figures are only estimates, of course, and are liable to normal statistical error. Yet even when we use conservative figures, the rate of background extinction through most of geological time is astonishingly low: *One to five species go extinct in any single year*. It is shocking to compare this very low rate of background extinction with estimates of extinction rates today. These, of course, are equally prone to error. Nevertheless, even conservative estimates of current extinction rates are so much higher than the calculated background rate that they seem to lose meaning: Conservative estimates of extinctions taking place in the tropical forests of the world today range between 4000 and 14,000 species per year!

Some of our best information about extinction comes from historical records, which have been maintained with any reliability only since about A.D. 1600—and for that interval only for a restricted suite of large, conspicuous organisms, including birds and mammals. Since 1600, 113 species of birds and 83 species of mammals are known to have gone extinct. Far more, about which we are only now extracting information from the fossil record, went extinct in the previous millennium. Around three-quarters of these extinctions took place on oceanic islands. Historical records also suggest that since 1600, extinction rates for these two groups have increased by a factor of 4, to produce the current extinction rates of around 0.5% of extant birds and 1% of mammals per century. Extinction rates in other groups of organisms have only begun to be tracked in this century, but the rates are alarming. In the United States, twice as many species of fish (350) were classified as endangered in 1996 as a decade earlier.

The major factor driving species to extinction in North America (and elsewhere as well) appears to be habitat destruction; any kill curve for the modern day should thus plot amount of habitat destruction against number of species going extinct. But this is not as straightforward as it seems. It turns out that the best way to understand species disappearance rates is through the use of *species–area curves*, where the number of species found in a given region is plotted against habitat size. The simplest of several equations to describe this relationship is

$$S = cA(z)$$

where S is the number of species, A is the area, and c and z are constants dictated by biological conditions. When the area and number of species are converted into logarithms, the curve can be plotted as a straight line whose slope z is given by

$$\log S = \log c - z \log A$$

From these simple formulas, very powerful mathematical generalizations emerge. For instance, the larger the size of a forest that is being reduced by clear-cutting or burning, the greater the number of species endangered. A salient point, however, is that the extinction of species established within a forest does not necessarily take place at the same time as the forest's destruction. Although many species are immediately eradicated, others, composed of small numbers of individuals, can hang on for extended periods, giving the impression that they have avoided extinction.

This is the central fallacy of today's conservation practices: No sort of environmental destruction, be it a meteor from outer space, the destruction of a forest, or the introduction of an Ice Age human hunter into a native ecosystem, causes immediate extinction of all the affected species. Some die out immediately, of course, but others hang on. They are doomed—but the extinction is protracted. There is what we might call an *extinction debt:* Decades or centuries after a habitat perturbation, extinction related to the perturbation may still be taking place. This is perhaps the least understood and most insidious aspect of habitat destruction. We can clear-cut a forest and then point out that the attendant extinctions are low, when in reality a larger number of extinctions will take place at some time in the future. We will have produced an extinction debt that has to be paid. This debt can also be incurred by other means, such as excessive fishing or hunting. We might curtail our hunting practices when some given population falls to very low numbers and think that we have succeeded in "saving" the species in question, when in reality we have produced an extinction debt that ultimately will have to be paid in full. Whales come immediately to mind. We have hunted many of the larger baleen whales nearly to extinction, but the well-publicized, 11th-hour moratorium on most commercial whaling seems to have saved most species from extinction. Or so we think. It is not certain that we have really achieved this success. A millennium from now we will know better how successful our "Save the Whales" campaigns have really been.

The various insights from species–area curves are not applicable only to the modern world. In his superb book *Conservation and Biodiversity*, ecologist Andrew Dobson of Princeton University makes the following point: "If humans have one ecological feature that distinguishes them from all other species, it is their ability to alter the landscape in which they live." Dobson was speaking of humans in the modern day, but his insight surely

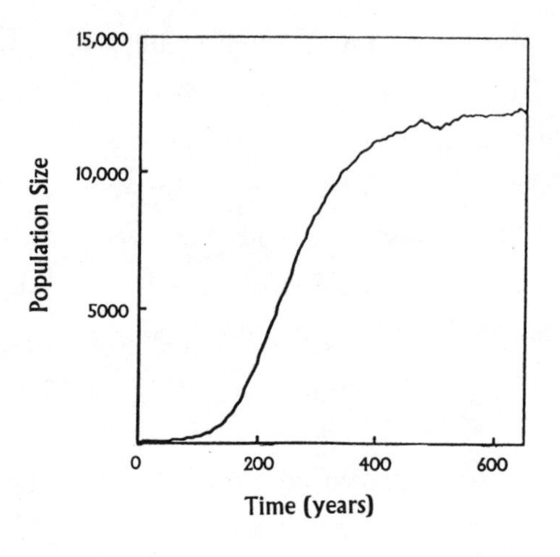

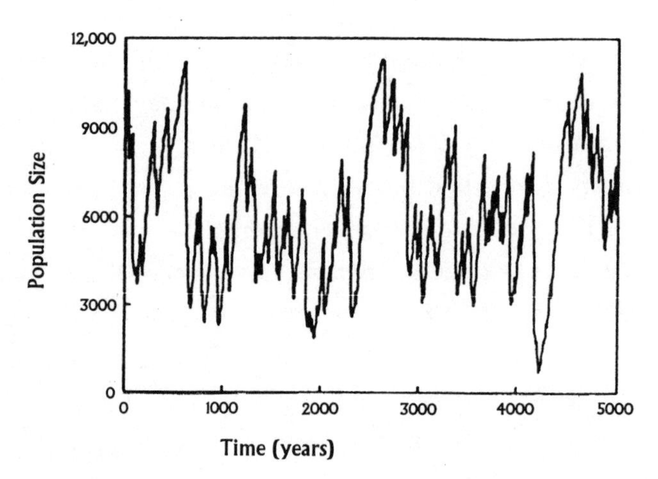

Two growth curves, one showing theoretical increase of an elephant herd under ideal conditions; the second simulating population size fluctuation in an elephant herd over 5,000 years with droughts of 10, 50, and 250 years interspersed.

applies to our distant ancestors; Landscape alterations is not a characteristic only of recent times. The archaeological record suggests that in every case where humans have arrived in some new habitat, landscape alteration has unfailingly occurred, through agricultural practices (such as slash and burn), through the rampant use of fire, through the introduction of nonnative species, or (even more indirectly) through changes brought about in the suite of animals present by the hunting or raising of livestock, which themselves alter the vegetational regime in any given area.

. . . .

The concept of the kill curve was originally adopted for tracking the fate of large suites of species, but it can also be used to track and examine the fate of individual species. Conservation biology has long used quantitative models to assess risk of extinction for individual species and to model the rise and fall of population numbers within species. Similar models are used in fisheries and game management. One of the goals of all such studies is to determine the minimum population size required to keep any given species "safe" from imminent extinction, or even to take that species out of the pool of "extinction debt" victims.

The number of individuals in any population is always fluctuating. There may be long-term trends toward increase or decrease, or even toward constancy, but such trends are themselves made up of shorter-term fluctuations. These fluctuations have traditionally been thought to be related to environmental factors: change in food supply, increased or decreased predation or competition, or alterations in the physical environment such as long-term temperature change. To understand these changes, ecologists have developed a series of equations that describe how birth and death rates—the ultimate determinants of population size—are affected by the external environment.

The earliest models used by ecologists to describe population change in any given species were called *deterministic models*: They assumed that birth and death rates remain constant. In these models, populations will always increase in size if birth rate exceeds death rate. Yet these models, though effective in predicting short-term changes, are usually doomed to failure because of a central omission: They do not take into account rare catastrophes and unforeseen events that can affect population size, such as the introduction of a new predator, a particularly violent storm, or a harsh winter. They also fail to reflect the fact that reproductive rates themselves are not constant. Andrew Dobson argues that any mathematical model that tries to account for population size must recognize three sources of variability: *demographic stochasticity*, which arises from chance events in the birth and survival of discrete individuals; *environmental stochasticity*, which is caused by changes in the weather, food resources, and other features of a population's habitat and by such natural disasters as fire, flood, and droughts, which occur at unpredictable intervals; and *reduced genetic diversity*, whereby the population becomes less able to adapt to pathogens or other adversity. Models that take these elements into account are called *stochastic models*.

Stochastic models differ from deterministic models in several important respects. They are more complicated and do not readily yield an actual size for a given population after a known amount of time; rather, they yield a probability that the population has reached a given size. Yet they have the great advantage of taking into account chance events that the deterministic models neglect.

Stochastic models have been commonplace tools in ecology for several decades. One of their most important implications is that any species has an *extinction threshold*: a minimum population below which the species is unlikely to survive. Small populations are vulnerable to extinction even if, on average, they are showing an increase in numbers (births outstrip deaths), because chance events can easily destroy the entire population. This finding also makes clear an ancillary threat: When the habitat of a population is split, thus dividing the species itself into many smaller fragments, the species is much more prone to extinction because of the effects of the extinction threshold. Just such an event appears to have occurred at the end of the Ice Age. The largest single habitat of the mammoths—the so-called mammoth steppe, a huge region stretching virtually the length of North America south of the ice sheets as well as through much of Europe and Siberia some 15,000 years ago—was split into many isolated pieces when the last glacial interval came to an end. The various populations of mammoths were also split into many tiny populations.

Stochastic models have greatly improved our ability to model nature and the population swings of local species. But are these models accurate?

One of the great mathematical revolutions of this century, recognition of the process known as chaos, suggests that even these later-generation models leave much to be desired.

The importance of unpredictable effects on populations was perhaps first appreciated by the great Oxford mathematical ecologist Robert May. In the 1970s, May showed that population fluctuations in many animal and plant species are not necessarily random but instead may be an aspect of chaos. In this case, although governed by precise mathematical rules, the behavior of a chaotic system is virtually impossible to predict. It may be that some populations of organisms can show wild fluctuations caused not by external conditions, such as climate change, but by deeply rooted and complex dynamics *within* the ecosystems in which they reside. In addition to showing that population size can be governed by chaotic properties, May also showed the *geographical distribution* of organisms may be affected by factors that are not necessarily related to the external environment. May and his colleagues showed that the variation of populations within a patchy (irregular) distribution may not be related simply to the favorability of each patch. It may be far more complex.

All of these findings have profound implications for conservation biology and for the understanding of mass extinctions. In their 1996 book *The Sixth Extinction*, Richard Leaky and Roger Lewin point out that

> the world of nature is not in equilibrium; it is not a "coordinated machine" striving for balance. It is a more interesting place than that. There is no denying that adaptation to local physical conditions and such external forces as climatic events helps shape the world we see. But it is also apparent that much of the pattern we recognize—both in time and space—emerges from nature herself. This is a thrilling insight, even if it means that the work of conservation management is made more difficult. It was long believed that population numbers could be controlled by managing external conditions (as far as possible). This must now be recognized as no longer the feasible option it was imagined to be.

Our mathematical models are but crude stepping stones along the path of ecological understanding. Yet for all their failings, they are at least showing us that there is a path.

. . . .

If modeling the fates of modern ecosystems and of the species within them is so difficult, can we ever succeed at the even more complex task of reassembling *ancient* communities? Such communities are made up of many extinct organisms and many more "invisible" organisms (those not preserved in the fossil record and therefore not amenable to any of our analyses) that surely participated in the economy of the ancient ecosystem we study. Those of us who study the past have learned to approach complex analyses with caution. Yet insights can be gleaned—in some cases, very powerful insights. Just such a breakthrough has occurred in the study of mastodon and mammoth populations.

As we have seen, population fluctuations have many causes. But what happens when some new type of predator appears in a habitat and suddenly begins reducing the size of previously unaffected target populations? This situation occurs in nature all the time, with the evolution or introduction of a new predatory agent or even a new disease. Ecologists have developed a series of models to show how such a new agent will ultimately affect the prey. These curves have long been used by fisheries scientists and game managers to document the effect of human fishing or hunting on known resources. They are now being used to study the effect of human hunting on mastodon and mammoth stocks of the late Pleistocene.

The first use of such models for ancient elephants was attempted by Paul Martin and one of his colleagues, J. Moisimann, in 1975. They tried to show how hunting of some proportion of the mammoth or mastodon population would affect overall population size using a model based on the extinction of mammoths in Europe. The Moismann and Martin model, as it came to be known, had three goals. The authors wished to know how many hunters might be sufficient to cause an extinction; how much time might be required for the extinction to occur; and how much evidence might be left in the fossil record.

The M&M model, as I will call it here, relied on various assumptions and on seven variables. These had to do with the initial mammoth biomass (mammoths per square mile of territory); the mammoth carrying capacity, a measure of how many individuals any given area could support; the rate of "replacement" of the mammoths by reproduction (or the annual growth rate of the herd); the initial human population size; the human population growth rate; the human carrying capacity, in people per square mile; and the prey destruction rate (the number of mammoths destroyed per person per year). The simulation yields radically different results depending on the values that are selected for the variables. It was from this work that the concept of the "front" (described in Chapter 6) emerged.

Objections to the M&M model were soon raised; they were mainly related to the values plugged in for the various variables. The most sensitive variables were the actual human population density, the prey destruction rate, and the human population growth rate.

By the late 1980s and early 1990s, the M&M model had been improved through better assumptions for the variables (derived from new data) and through slight changes to the model itself. In 1989, for example, Whittington and Dyke published the results they got by using an improved version of the M&M model. They reached very different conclusions from those of Moisimann and Martin's. Whittington and Dyke showed that because so many variables are being used in the M&M model, there are a variety of threshold values above which extinction takes place. They also showed that there is no single way to "solve the equation" that results in extinction for a given prey species. Very slight differences in these variables may result in survival for one species and death for another. The authors point out that this insight alone seems to answer a reservation expressed in 1943 by Loren Eiseley, who questioned the overkill hypothesis on the grounds that some New World mammal species went extinct whereas other, closely related forms did not. Very small variations in the biomass replacement rate can mean the difference between survival and death. Many species may be barely surviving without external predation; add extra stress and they quickly slip into extinction. The rate at which a species can reproduce is also a critical factor in who lives and who dies. Because mammoths and mastodons apparently had an even lower reproductive rate than modern elephants, they were highly susceptible to predation—not in the sense that they were easily killed by predators, but in the sense that even slight amounts of predation would quickly imperil the entire population. In other words, just as we might have expected, extinction can come about for a variety of reasons, and the variables we must examine are several and interlocking.

More recent computer modeling indicates that it is not necessary to assume an expanding human "front" of big-game-hunting Paleo-Indians (such as the Clovis people) to account for the extinction of the Ice Age megamammals. A Russian named Budyko, trying to understand the extinction of the mammoths in Siberia and Northern Europe, used equations and modeling to demonstrate that an extinction that has been supposed to have been caused by a single event, such as climate change, could actually have been caused by long-term human predation. Budyko argued that a human population with very slow growth—just the opposite of the assumptions in the Moismann and Martin model—could have been preying on mammoths and mastodons for tens of thousands of years with little apparent effect until the human population reached some threshold density. After that, sudden and massive extinctions occurred. The prey population crashed to very low levels. Describing Budyko's work, Whittington and Dyke write, "His well-explained model demonstrates the possibility that humans alone could have been the agents of megafaunal extinctions."

There are many examples of this in case histories of fisheries, where large, supposedly extinction-resistant stocks of fish suddenly crash when the fishing rate is enhanced just slightly. Some threshold is surpassed, and the stock suddenly falls to very low numbers. The population crash of anchovies off California in the 1940s and that of king crab in Alaska in the 1980s illustrate this effect. The extinction of the passenger pigeon in the early 1900s may be another example. The huge populations of this bird had been preyed on by humans for decades. Yet when some critical predation level was reached, the seemingly endless flocks quite suddenly vanished. The mammoths of Siberia and North America may have suffered the same fate.

The most recent mathematical model that explicitly describes mammoth extinction was developed by Stephen Mithen of Reading University. Mithen's program, which the user initiates by typing "Lets hunt mammoths!" on the keyboard, allows one to manipulate parameters such as population size and kill rate. A new wrinkle in Mithen's program, however, is that it shows how hunting affects populations already under stress. The previous models show the effect of a burgeoning human population. Eventually some threshold limit is exceeded, and the prey population crashes. Mithen's models can incorporate the assumption that hunting intensity increases in a small human population. In this scenario, if early hunters killed more than 2% of the total mammoth population in any given year, the entire mammoth population would be condemned to extinction—eventually. The important aspect of this computer model is that it explicitly shows the effects of an "extinction debt." Killing *over* 2% of the population (each year) does not cause an immediate population crash or even an extinction in a century, but it inevitably consigns the hunted population to extinction several centuries down the road. When predation occurs at just slightly more than the maximum allowable culling rate, the inevitable loss of the entire population may take as long as 400 years.

The computer models produced by various scientists suggest that neither climate alone nor hunting by the early humans was sufficient to kill off the mammoth and mastodons herds. However, these new kill curves suggest that shrinking and newly isolated biological ranges in deteriorating environments, coupled with hunting, were indeed sufficient to cause a protracted and ultimately catastrophic decline in the great elephants. The extinction did not occur across a huge population, but among many tiny populations no longer in contact. But what proof is there that the elephants *were* in decline—that environments were deteriorating and habitats being isolated? Is there *any* evidence of environmental stress sufficient to cause such a decline in the megamammal populations that they became vulnerable to even slight hunting pressure? Yes. Important clues to this long-running murder mystery can be gleaned not only from the fossil record but also from the elephant killing fields of modern Africa.

. . . .

I have always had a certain vision of Africa: vast savannas filled with game, snarling lions defending freshly killed zebras from packs and flocks of scavengers, charging rhinos and sky-scraping giraffes; sluggish streams infested with scaly crocodile, and above all, elephants. Elephants in vast herds flapping wing-like ears as they graze beneath the brooding Mt. Kilamanjaro. Such places do exist in Africa still. But by and large, this vision describes the national game parks of Africa; like most places on Earth, Africa has very little wildlife left outside the parks.

There still exist great wild regions—the deserts of Africa are as remote and lonely as ever. But areas that can support large game are also fit to support large human populations, and in spite of starvation, drought, AIDS, and other scourges, the human population in Africa continues to swell. Over most of the continent, each human mother produces at least four babies in her lifetime, and in some places, such as Kenya, that number approaches six. The result is that most of the arable land in Africa has been cultivated—and must be cultivated if mass human starvation is to be avoided. Crops are the only hope for feeding all of the hungry mouths, and crops do not do well when they are situated in the path of wild elephants.

Elephants are much like the dinosaurs of the past: They disrupt plants through their feeding and their movement. Each elephant eats enormous quantities each day, but even more deleterious to floral health in any region inhabited by elephants is the effect of their daily marches. Such large animals trample small and even medium-sized plants. Elephants moving through cultivated land spell disaster. No wonder, then, that elephants and the growing population of African farmers have been on a collision course for decades. To make matters worse, selling ivory from mature elephant tusks (though doing so is illegal) can yield money much faster than raising crops. Finally, you can *eat* a slain elephant, and modern rifles make the task of killing them far easier than that faced by the ancient Clovis people.

We can see the result of this conflict simply by shifting our gaze from Africa to India. There the wild populations of *Elephas*, the Indian elephant, have been extinct for years. Many elephants survive as beasts of burden. But they do so at very low population numbers, and never again will herds of the Indian elephants be seen in the wild in large numbers. Is this the fate of the African elephant as well?

It has long been recognized that the survival of the African elephant is threatened by crowding as well as poaching. The last hope for these elephants is now assumed to be the great game parks. But even there the elephants are endangered, perhaps in part because of the very success of the parks in reducing the rate of poaching, which decimated the overall populations of African elephants in the 1970s. Recently, to keep the great herds of elephants from destroying the parks through overpopulation and then "escaping" from the parks to return to their native feeding areas (which are now the sites of human farms), elephants in the game parks have had to be periodically "culled." In other words, killed. It is from these culling episodes, however, that we have gleaned some of our richest and most pertinent information concerning *taphonomy*, or the way elephant bones go from the living animals to the fossil record. One scientist, especially, has recently transformed our understanding of the elephant graveyards and, in so doing, has opened up a new world of reinterpretation of graveyards past.

. . . .

I cannot imagine a sadder experience than watching elephant die-offs: times when large numbers of elephants are killed by severe environmental change, usually drought. But even more gruesome would be to witness elephant culling, when elephants are deliberately slaughtered, not for their meat or tusks, but simply because there are too many of them to live in the small preserves to which we have sentenced them. Yet the insight that such episodes of death may yield about the eventual fate of the carcasses and bones can help us interpret the fossil record.

Anthropologist Gary Haynes of the University of Nevada has repeatedly visited the elephants graveyards and killing fields, and his experience resulted in a landmark book, *Mammoths, Mastodons and Elephants*, published in

1991. Haynes began to study the ways in which modern elephants die in an effort to understand whether fossil elephant bone assemblages were caused by human or nonhuman factors. Did the Pleistocene mammoths and mastodons die off, or were they killed off? The answer to this question may be found in the sites where elephants naturally die off, around shrinking water holes in southern African game preserves, in times of drought or other environmental stress.

Because they travel in herds, modern elephants are relatively invulnerable to predation. Young males may be killed when they wander off, and very young elephants may be killed during times of environmental stress that weakens protective herding behavior. But for elephants, as for us humans in the 20th century, predators are a minor worry. Like us, the elephants have two major enemies: drought and humans. Droughts, which reduce both water and food supplies, can take a tremendous toll on elephants in the wild. During especially severe episodes, up to 20% of a herd may die off each year. Coupled with elephants' low reproductive rate, this would suggest that they are very prone to extinction. Apparently, however, they are not. Haynes's observations have convinced him that it would take extraordinary conditions to drive modern elephants into extinction:

> Yet in spite of recurring heavy mortality during drought, *Loxodonta* [the African elephant] populations have been able to bounce back and maintain healthy growth rates. This ability to recover following serious die offs in Africa is striking and has made me think again about the effects of Late Pleistocene climatic stresses on mammoths and mastodon. . . . The most important lessons to be learned from studies of modern elephant die offs have to do with the proboscidean responses to climatic change and with proboscidean resilience allowing recovery from nearly any environmental stress *except human overhunting* [emphasis added].

Drought is perhaps the greatest "natural" killer of the elephants. Because much of their African range is characterized by dry and wet seasons, and because the amount of rainfall can fluctuate from year to year, the elephants' migration patterns are largely related to finding water, for where there is water, there is food. Elephants also need great quantities of water to help in digesting food and to regulate their body temperatures, especially in very hot conditions. Although adult elephants can go without water for several days, this is clearly not to their liking. During dry seasons, elephants will dig "wells" up to 3 meters deep into drying water holes. These sites become one type of elephant graveyard.

Haynes' studies were concentrated around these water "seeps"—springs and water holes that are nearly dry during the hottest months. In drought years, these regions become the sites of natural elephant die-offs. Smaller elephants were observed to die nearest these seeps, and often their carcasses fell within the elephant-dug "wells." Over time, these skeletons were disarticulated and crushed as more elephants came to the wells. Because of the water content in the regions, however, the seep and well sites also have a high potential for fossilization of the elephant bones. These sites become natural bone accumulations, and as the elephants come back to the same sites year after year, the bones that slowly accumulate in these sites become a rich, long-term record of elephant deaths.

The transformation from living animal to discovered fossil requires a string of highly unlikely coincidences. Water is almost always a prerequisite for fossilization. Animals that die on dry plains have very little chance of being buried and preserved in sediment before they are completely destroyed by scavengers and decomposing organisms. In water-rich mud, however, the bones can quickly become inaccessible to scavengers. This may be one reason why hominid fossils are so rare: Our ancestors seem to have lived on dry plains, where their bones were quickly scavenged. So too with the many elephants that die away from water.

The types and shapes of bones found around the seep sites in Africa have yielded fascinating clues to how and when elephants die. They also yield valuable insight into the past. Because Haynes and others have accumulated so much information about modern elephant kill sites, they have a data base with which to compare bone "profiles" from mammoth and mastodon bone accumulation sites.

Gary Haynes saw more than natural elephant die-offs. In the early 1980s, because of successful antipoaching programs and the increased numbers of farmers in the area, the country of Zimbabwe in southern Africa decided that it had too many elephants. Between 1983 and 1986, the Zimbabwe Department of National Parks shot 9000 elephants, leaving the carcasses where they fell. Haynes, who witnessed much of this carnage, was able over several years to watch the progression from living elephant to disarticulated bones.

This "actualistic" work—and long observations by Haynes and others—have provided our best information yet on how to interpret fossil elephant sites. Several generalizations have emerged. First, it is highly unlikely that the bones of a mammoth or mastodon killed by an Ice Age hunter would ever be incorporated into the fossil record unless the kill site was near water. Single elephants that die in Africa today or are killed by poachers will not, in all probability, enter the stratigraphic record—and *will not become fossils*—without rapid (geological) envelopment by sediment. The same must have been true of Ice Age elephants. Unless the mammoth hunters killed their prey near water (lake, stream, swamp, or bog), it is highly unlikely that any evidence of the killing would have been preserved in the fossil record. This seems to be borne out by the paucity of mammoth kill sites associated with human cultural artifacts. In North America, for instance, only about a dozen sites with both mammoth bones and Clovis artifacts are known. Second, the rich fossil record of mammoth and mastodon bone accumulations consists largely of die-offs that occurred not through the agency of humans hunters but through the same mechanisms operating in Africa today—die-offs near ephemeral water.

These observations have enormous relevance to interpretation of the role humans have played in the extinction of mammoths and mastodons. The well-known mammoth bone huts and mammoth bones beds in Central and Eastern Europe and Asia have long been considered to be the sites of mass herd kills. Yet these bone accumulations show age profiles completely consistent with the elephant die-offs in Africa. Nevertheless, the image of Stone Age mammoth hunters is a cherished cultural icon.

Gary Haynes has had the temerity to burst this bubble: "the dramatic image of mammoth-hunting specialists living during the Upper Paleolithic in the cold steppes of northern Eurasia has become firmly established in world archaeological thought. It is an image based on a very selective example. The current folklore about mammoth hunters is an imaginative interpretation that lacked a balanced perspective." Early humans knew about the graveyards of the fossil elephants, and they knew about the ancient ponds and seeps where huge piles of bones and tusks could be found and put to use. They probably also learned that at such sites, small proboscideans, those most at risk, could occasionally be picked out of the herds with little danger. Our ancestors are not likely to have killed great herds or even great individuals. We probably camped around the elephant congregation sites and picked off small numbers of stragglers and young. But even small numbers of kills may have been enough, over time, to nudge these great creatures onto the road to extinction.

The work of Gary Haynes is the most comprehensive examination of the way elephants die and of how the resulting bone accumulations compare with the fossil record of extinct proboscideans. Haynes also notes interesting similarities and differences between the record of ancient proboscideans in the New World and that in the Old. Clearly, humans had a far longer association with elephants in the Old World, and the extinction of these great beasts seems to have been complete in many regions before humans arrived in North America. Mammoths were extinct in Europe and China before humans ever reached the New World.

It is also clear that great accumulations of ancient elephant bones were forming in Europe and Asia between 18,000 and about 14,000 years ago, which indicates that environmental conditions *were* changing enough to cause great die-offs of these beasts. There can be no question that the elephant herds of the Ice Age were under severe environmental stress during the last glacial interval and that their numbers were being reduced. It is in this crucial respect that I believe the Ice Age extinction differs from the End Cretaceous extinction: There is, as yet, no convincing evidence that dinosaurs were undergoing die-offs or even reduction in species diversity during the last 10,000,000 years of their reign, let alone the last 10,000.

Does this mean that climate change is guilty of these great extinctions? It is here that the modeling of overkill by computers and the recognition of extinction debt yield crucial insights. For so long, we have believed that any species exterminated by humans would have disappeared rapidly and that humans would have been responsible for the preponderance of deaths. But that is clearly not the case here. Only a slight human-caused increase in the natural level of mastodon and mammoth deaths would have been sufficient to bring extinction. Perhaps we do not need to invoke overkill. "Anykill" will suffice. Haynes has made the same point:

> A sweeping conclusion that would indisputably point the blame for the proboscidean extinction on Clovis hunters would be a literary triumph but a scientific impossibility. The ultimate impact of the Clovis culture is difficult to gauge. Even before the Clovis point and associated traits appeared, late glacial mammoth and mastodon populations probably were in overall decline throughout the last few millennia of the Pleistocene, although some intervals would have been more favorable than others. . . . But Clovis hunters perhaps made an already bad situation worse for megamammals: while climatic changes were driving proboscideans to die off, the added stress of Clovis hunting drove them to *die out*. In the absence of Clovis hunters, mammoths and mastodons (and other megafauna) should have been able to survive the changing conditions of terminal Pleistocene environments.

Haynes summarized his view of the human role in the Ice Age elephant extinction as follows:

> I propose that humans opportunistically spread their worldwide range in northern regions only after they had found a resource (mammoths) that was becoming more and more vulnerable because of environmental stress such as harsher winters or seasonal drought. In other words, late Pleistocene hunting and gathering groups were not mammoth specialists who spread into the New World and caused the disappearance of mammoths because of their superiority as big-game hunters. Rather, the rapid spread of humans was in response to their awareness that mammoths were clustering in certain identical regions and habitats, where die offs were resulting. If people did indeed come to the New World because they discovered that game animals were easy prey for part of the year, those people may have reacted to the growing scarcity of their preferred subsistence base by increasing their mobility, to seek out the dwindling localized big game populations. Greater mobility would have led to further explorations and rapid colonization of unoccupied ranges where mammoths and perhaps other large animals still survived. (Haynes, p. 284.)

If this view is correct, there is no need for Paul Martin's concept of the front. The Clovis indeed would have spread southward through the continent but would have backtracked considerably seeking out elephant die-off sites where the young and lame could be easily killed.

. . . .

Long ago, in Africa, the magic cauldron of evolutionary change created elephants and humans. Both departed the continent of their origin and wandered to the far corners of the earth. Somewhere during that journey, one became prey, the other predator, as both were caught up in the climatic convulsions we call the Ice Age. The most recently discovered clues to this ancient puzzle give the clearest view yet of how convulsive the changes wrought by the Ice Age really were and to what degree they affected human history and that of the mammoths.

Elephant hunt. (From The Life and Explorations of David Livingstone, LL.D.)

8

The Lost World

EVERYONE HAS SOME FAVORITE BOOKS that are read over and over until they are worn out. I have many. But among my very favorites, which I can reread without any fatigue, is Sir Arthur Conan Doyle's: *The Lost World*. The thesis of Conan Doyle's book is simple: Deep in the middle of the Amazon jungle is a high plateau still populated by prehistoric creatures. On a dare, our narrator (one Edward Malone) accompanies Professor Edward Challenger, Lord John Roxton, and Professor Summerlee into the great South American wilderness where they encounter in all its saurian majesty, high atop this ancient plateau, a place where the last remains of the Mesozoic are entrapped. It is one of the enduring scientific fairy tales, that great prehistoric beasts have escaped extinction in a remote refuge. The Loch Ness monster, the Sasquach, the pygmy brontosauruses reported in Africa in the last decade—all reflect this whimsical hope that somehow, somewhere, some great creatures from the antediluvian past live on in isolated prehistoric splendor.

Doyle first published his fable in 1912, probably about the last time when such a tale could be remotely believable, for even by that time most of the world had been thoroughly mapped and explored, and there existed very few places where a great ark of prehistoric beasts could still be hiding. In our day of ecotourism and orbiting surveillance satellites, such a tale is utterly unbelievable. Yet sometimes we are indeed surprised by the unexpected discovery of some ancient creature, small and unexceptional to anyone save some very excited boffin. And even more rarely, something *bigger* stumbles into view, witness the discovery in 1938 of the coelacanth (a giant blue fish thought to have been extinct since the Mesozoic but dredged up off South Africa) or the more recent find of Jurassic-aged pine trees in Australia. For the most part, however, the "living fossils" are more nondescript: a small bug or deep-sea worm. Yet the concept of a refuge is an important one for our study of the Ice Age. What if we knew a place where the same *vegetation* eaten by the mammoths and mastodons could be found, a lost world where these great proboscideans actually held out for thousands of years after their extinction everywhere else on Earth? It turns out that several such places existed until relatively recently, and the discovery of these Ice Age lost worlds has yielded some important insights into the mystery of who—or what—caused the extinction of the great Ice Age mammals. We can inquire not only why the mammoths or mastodons survived in small refuges in a few places on Earth, and nowhere else, but also why they eventually died out even in these refuges.

. . . .

High in the Arctic Circle, separated from the Siberian mainland by 100 miles of ice-strewn sea, Wrangel Island is a bleak, frozen wasteland that has proved inhospitable to humankind. Its original human inhabitants were hardy Aleuts, and the descendants of these pioneers live there still, eking out a meager existence on seal, bear, and whalemeat. But for the last several decades, many of the human residents of Wrangel Island have been scientists who have come to study this modern-day lost world, a botanical refuge of the great Ice Age. Unlike other nearby islands, Wrangel has retained a flora virtually identical to that which covered much of northern Europe and North America during the Ice Age. Each brief, Arctic summer, a diverse assemblage of grass and low shrubs emerges from the permafrost. These plants belong to species known to have been food for the great Ice Age mammoths and mastodons.

Wrangel Island is more than an Ice Age plant museum, for it long served as a lost world for prehistoric animals as well as plants. A 1993 study confirmed that Wrangel Island is the gravesite of the last mammoth to have walked our planet. Mammoth bones are not rare elsewhere; in nearby Siberia, for instance, the tusks and bones of mammoths and mastodons are so abundant that the native peoples find new specimens each year. But most of these bones come from creatures that lived more than 12,000 years ago, and in no case less than 9500 years ago. On Wrangel Island, mammoths and other relicts of the great Ice Age lived as recently as 4000 years ago, in a "Pleistocene Park" that existed more than 5000 years longer than anywhere else on Earth. Mammoths were still living on Wrangel Island when the first pyramids were being built in Egypt.

Wrangel is not unique in having been a sanctuary for the Ice Age megafauna long after these great beasts died out everywhere else. A few other refuges are known, and all these ancient sanctuaries were islands. Did something about the microclimate of various islands stave off the grim reaper of extinction? Or were they preserves simply because ancient humans of the time were not very efficient sailors?

Some of the most spectacular preserves for ancient mammals seem to have been in the Mediterranean Sea. The preservation of *any* larger mammals there is all the more striking because the Ice Age megafauna in Europe appears to have gone extinct even earlier than that of North America. For instance, Paul Martin has shown that the last mammoth in Europe is dated at slightly more than 12,000 years ago, whereas the last Ukrainian mammoths are about 13,000 years in age. The woolly rhinoceros died out about 11,000 years ago in Europe; the bison lasted until 10,000 years ago. It appears that mammoths were vanishingly rare in Europe at a time when they were still numerous in North America.

While the larger mammals of Europe were gradually dying out or were gradually being killed off by the slowly burgeoning human population of the Paleolithic, a very different evolutionary scenario was unfolding in the Mediterranean Sea and its larger islands. Even with the rise and fall of sea level during the Ice Age, the larger islands of the Mediterranean were fully cut off from land. Occasionally, larger animals reached these shores by swimming or rafting, and over the long course of the ice ages, separate populations of new species were established. Curiously, in almost every case, these various mammalian species underwent dwarfing. Generation after generation, smaller size seems to have been favored, and by the end of the Ice Age, each island had a distinctive assemblage of dwarfed creatures. Thirteen endemic extinct genera are known from the Balearics, Sicily, Malta, Sardinia-Corsica, Crete, and Cyprus, including elephants, hippos, and varieties of deer. All seem to share a common trait, a trait also exhibited by the mammoths from Wrangel Island: All are dwarfs. One older form from Sicily is the smallest adult elephant ever to evolve. It was the size of a medium-sized dog, tusks and all. The Wrangel mammoths were larger, being about 6 feet high at the shoulder, in contrast to the normal mammoth size of 10 to 11 feet high.

The dwarf mammal faunas from the Mediterranean were long thought to pre-date the end of the Ice Age, and in many cases the oldest examples of these species indicate that this curious assemblage of mammals lived on these islands for hundreds of thousands of years—without humans. It is the more recent dates that are of special interest, however. A new form of age dating via changes in bone protein suggests that the youngest fossil bones of the dwarf elephants date to only about 5000 years ago, long after the end of the Ice Age. If climate change killed off all the larger animals along the shores of the Mediterranean more than 12,000 years ago, why did the creatures survive so much longer on these islands? Was it simply because the island refuges were separated from the mainland—and thus from humankind—by the protective sea?

. . . .

I have this vision of the end of the Ice Age: gradually warming temperatures, the barely perceptible retreat of the ice (it is not for nothing that we label the slowest imaginable movement "glacial"), and summer weather lengthening slowly. The changes took millennia, occurring so slowly that no human cultural history might even take note of them, except in ancient campfire tales. My vision is that of a world of calm cold during the Ice Age replaced by the more tumultuous and stormy era of the present day, for warming means increased storm activity. In 1993, however, that view came tumbling down, all on account of new observations derived from the great glacial storehouse of Greenland.

The very name is a cruel joke, for there is precious little green on this ice-covered island. For millions of years Greenland has been a giant refrigerator, a place of howling winds and glaciers, a land whose greatest claim to fame among geologists and paleontologists had been its storehouse of fossils from the Age of Coal of 300 million years ago. But during the late 1980s and early 1990s, another type of fossil record was for the first time extracted from the wintry island, and not from stratified rock but from stratified ice. Greenland carries one of Earth's greatest reserves of fresh water locked up as ice—ice that has been deposited, year after year, for more than 2000 millennia. It is from this ice that scientists have recently uncovered one of the most startling discoveries of the 20th century, which may help solve two perplexing mysteries. Why did the discovery of agriculture occur so late in humankind's history on Earth? And why were late Stone Age societies of 25,000 to 15,000 years ago so culturally complex, yet so impoverished architecturally?

The answer to the first question has seemed evident enough: The invention of agriculture and the formation of villages and towns had to await the evolution of a sufficiently intelligent human species. According to this scenario, Paleolithic humans were not smart enough to have conceived of agriculture and architecture. Yet if *Homo sapiens sapiens*, the Moderns, evolved as current evolutionary theory demands, then something is clearly amiss. Humans equal in intelligence to any on Earth today have probably been on this planet for a minimum of 100,000 years. How many Einsteins and Newtons must have lived during our species's long existence, and why couldn't they figure out that putting a seed in the ground causes a plant, a food plant, to grow? Why did we—*Homo sapiens sapiens*—spend a minimum of 100,000 years (and as much as 200,000 years) living in the open or in caves, living at low population numbers, living by hunting and gathering, without benefit of anything but the most rudimentary technology, and most important, without agriculture? For at least 90,000 years, our forebears and intellectual equals seem to have stared stoically through campfires at predators and scavengers, cold and starvation. And then, about 10,000 years ago, the nature of life on Earth radically changed. As the last of the Ice Age megamammals went extinct, we as a species began to multiply and reach population numbers never seen before. Within a few short millennia we had begun to craft complex tools, to smelt metal, to domesticate animals, and to build villages and towns and finally cities. And most important of all, humans discovered agriculture at about the same time as the last mammoths and mastodons of North and South America died out. Large-scale agriculture first appeared in Europe and the Middle East about 9000 years ago and in East Asia 8500 years ago. What trigger event opened the door to agriculture and set the scene for a revolution in human lifestyle? Clues to this mystery seem to lie in the thick glacial storehouse that is Greenland.

Great scientific discoveries usually come from the most unexpected sources, and such was the case in 1993. After two decades of patient collection, followed by interminable isotopic analyses of the ancient, fossil ice, scientists from Europe and America finally completed their analysis of Greenland ice core samples dating back 200,000 years. They had expected to find a record of climate stability interspersed with epochs of temperature change, each coinciding with the advance and retreat of the great Ice Age glaciers. They found nothing of the sort. The numbers emerging from the great mass spectroscopes across the world showed that fluctuations of Earth's climate have been far more severe, and have occurred much more abruptly than any scientist had postulated—until 10,000 years ago, that is. This new discovery makes possible an entirely new interpretation of the rise of human civilization, and it certainly shows that our present-day weather—one of the prime bases for the concept of Uniformitarianism—is in fact very aberrant. We are currently in a state of calm, a period that has lasted 10,000 years. Before that things were anything but calm.

For much of the last 2.5 million years, crystals of ice in the Greenland ice cap have faithfully adsorbed minute quantities of oxygen and carbon isotopes, and in the process they have created a record of Earth's climate. By looking at isotopic ratios of oxygen, we can deduce ancient temperatures. The analysis of oxygen isotopes from the Greenland ice cores have shown that, contrary to popular scientific belief, the climate over the past 250,000 years has changed frequently and abruptly; the magnitude of the global temperature changes has been far greater, and their intervals far shorter, than anyone imagined.

Dr. J. White of the Institute of Arctic and Alpine Research at the University of Colorado noted in a recent summary of the project that between 200,000 years ago and 10,000 years ago, average global temperature had changed as much as 18°F in a few decades. The current *average* global temperature is 59°F. Imagine that it suddenly shot up to 75°F or sank to 40°F in a century or less. Another of the researchers working on this problem, Dr. Minze Stuiver of University of Washington has told me that such dramatic changes could have taken place in as little as 5 *years*. We have no experience of such a world; such sudden perturbations in temperature would enormously affect the atmospheric circulation patterns, the great gyres that redistribute Earth's heat. At a minimum, these sudden changes would create catastrophic storms of unbelievable magnitude and fury. Yet such changes were common until 10,000 years ago. Imagine a world where storms that dwarf Hurricane Andrew lash the continents not once a century but several time each year, every year. Imagine a world where tropical belts are suddenly assaulted by snow each year. This was our world until 10,000 years

Asiatic deluge. Sudden global temperature changes over the past 250,000 years have been recorded; catastrophic storms of unbelievable magnitude and fury were common. (From L. Figuier, La Terre Avant Le Déluge, 1864.)

ago, when, according to the new studies from Greenland, a miracle happened: The sudden shifts in the weather stopped.

In 1993 it was discovered that 10,000 years ago, intense global weather changes that had been the norm for the past 2.5 million years suddenly disappeared; the weather entered a 10-millennium calm. Very soon after the start of this calm, we as a species began to build villages and then cities. We learned to smelt metal. And most important, we learned how to tame crops and domesticate animals. Human population numbers began to soar. Was the moderation in climate coincident with the extinction of the larger Ice Age mammals in North and South America, creatures perhaps adapted to change, not stability? With the cessation of change, did stable plant communities arise, the great forests spreading across the northern continents as climax communities? Or did the change from storm to calm, from cold to warm, from glacial to interglacial, occur after the great extinction was over?

Of one thing I am sure: There must be a connection between the cessation of mad temperate swings, 10,000 years ago, and the rise of human agriculture and civilization. And as we learned to sow and reap, surely our numbers rose as never before. With these changes our wandering and exploring may have even increased, enabling us to find the last regions of mammoth steppe, to sail the seas, to conquer the last great islands such as Crete, New Zealand, Sardinia, Madagascar, Hawaii, New Caledonia, Vancouver Island, and Wrangel Island—to seek out the remaining refuges of the great mammals or flightless birds, to find these lost worlds, and to destroy them.

ICE AGE FAUNA. (FROM L. FIGUIER, *LA TERRE AVANT LE DÉLUGE*, 1864.)

9

Nevermore

The seventh of March in the year 1995 was raw and blustery, typical for the cusp between the cold rain of winter and the somewhat warmer rain of spring in Seattle. The crocuses were up, the daffodils poised; the magnolias were showing signs that equinox would again occur this year, and I took solace in that. I remember being tired that night; my scientific partner and close friend, Bruce Saunders, was visiting from Bryn Mawr College for the week, and as usual we had been talking for hours each day, thinking about our favorite group of animals (the chambered cephalopods), debating, criticizing, insulting each other and our scientific competitors, berating the university administrators who complicated our lives, but doing what we like best—doing science. At each rapidly passing day's end we were flat; talking all day is as tiring as it is exhilarating, and whoever said that thinking is not hard work should try spending the day with Dr. Saunders. Nevertheless, although home and dinner awaited us, I suggested a detour, for a lecture.

We strolled up to the university's largest lecture hall, and I was stunned by the size of the crowd already assembled, even though we were early. Hundreds of people lined the entrance to the building, an almost unruly mob trying to find seats in an overflowing lecture hall. The lecturer I had steered us toward was Dr. Peter Raven, Director of the Missouri Botanical Survey and one of the first scientists in America to draw attention to the biodiversity crises confronting the great tropical rain forests. Although the environmental war cry "Save the Rain Forests" is now widely sounded, it was this man, several decades ago, who drew public and scientific attention to the problem of wholesale rain forest destruction, especially in the Amazon Basin. Raven, like Rachel Carson before him, was one of my heroes.

I was aware that Raven was well known, but I was both surprised and heartened that so many from the university community would turn out for a lecture that I knew in advance would be a bit of bad news. Finding seats was impossible, so I was sitting on the floor when someone quite unlike Peter Raven walked up to the podium. Oliver Sacks was introduced, and I knew we had made a terrible mistake. We marched quickly out of a room still filling, to arrive at a much smaller lecture hall next door. Here too an appreciable crowd had gathered, and the real Raven was just beginning his discourse on the state of the biological diversity in the world today.

Raven was as spellbinding as he was depressing. A kindly, white-haired man perhaps in his sixties, he spoke without notes all in a matter-of-fact tone that belied his grim message.

He began his talk succinctly. "We must take a more engaged position with respect to the current state of biological diversity on this planet." And with that, he began trotting out facts as I scribbled notes on the back of computer-generated cladograms I had been preparing with Saunders that day.

Raven talked largely about forests, clearly his passion. Forests have been a part of this planet for more than 300 million years, and although the species have changed over that long period of time, the nature of the forests has changed little. According to Raven, humankind has, over the brief moment of history during which we have been on this planet, treated the forests as though they were infinite, as indeed they must have seemed for much of our existence. The challenge facing us is to change that belief system—the ingrained conviction that all resources on this planet are inexhaustible—and acknowledge that they are quite limited, are clearly endangered, and must be viewed as at best sustainable.

The forests are Earth's great ark of species. Although the land surface of our globe is only one-third that of the oceans, it appears that 80% to 90% of the total biodiversity of the planet is found on land, and most of that occurs in tropical forests. As we destroy these forests, we destroy species. Raven estimates that 6 to 7 million species of organisms live in the tropical rain forests and that only about 5% of these are known to science. Because we have such a poor understanding of how many species really exist, it is next to impossible to derive hard data on how many went extinct in the last century or decade and how many will so do in next decade or century. Like most paleontologists, Raven believes that the world has attained the highest level of biological diversity ever in its history—and that the number of species on Earth will soon begin to plummet from this plateau.

There appear to be several forces driving a reduction in biodiversity—a *destruction* of biodiversity, to be less delicate. The most important of these seems to be the rapid increase in human population numbers. Ten thousand years ago, there may have been at most 2 to 3 million humans scattered around the globe. There were no cities, no great population centers; humans were rare beasts living in nomadic clans or groups or in settlements of little lasting construction. There were fewer people on the globe than are now found in any large American city. Two thousand years ago, the number had swelled almost a hundred-fold to between 130 million and perhaps 200 million people. Our first billion was reached in the year 1800. If we take the time of origin of our species as about 100,000 years ago, then it took us 100,000 years to reach the billion-person population plateau. Then things sped up considerably. We reached 2 billion people in 1930, about a thousand times faster than it took us to reach the first billion. And the rate of increase kept rising. By 1950, only 20 years later, we had reached 2.5 billion souls. By 1995 we hit 5.7 billion. There will be approximately 7 billion people living on Earth by 2020 and an estimated 11 billion by 2050 to 2100.

This recent ballooning of human population has completely changed the nature of problems on Earth, and it goes without saying that pressures on our planet's environment are completely unlike any exerted in its past.

It is not only the number of people on Earth that has changed, but also where they are found. In 1950 about one-third of the human population lived in what we euphemistically call "industrialized" or "developed" counties. In 1995 that fraction had dropped to about one-fifth, and it should drop to one-sixth by 2020. The population the United States represents about 4.5% of the total human population. However, the United States has an effect on the globe that is way out of proportion to its population. For instance, Raven has estimated that people living in the United States produce 25% to 30% of total world pollution. We control 20% of the global economy. Of the 3000 culturally and linguistically distinct groups of humans on Earth, those who call themselves "Americans" are, as a population, the wealthiest alive and the wealthiest in history. One consequence of this is that we consume more of Earth's resources than any other country. We also use more energy per capita than any other county, and it's not hard to understand why: since 1945, in America, the cost of gasoline has dropped by 33% in adjusted costs.

Much of the development of America has been at the expense of the rain forests. Rain forest conversion—a conversion that changes the forest first to fields and then, usually, to overgrazed, eroded, and infertile land within a generation—is perhaps the most direct cause of biodiversity loss. Raven reports that 25% of the world's topsoil has been lost since 1945, by which he means that it has been stripped from the surface and redeposited either in the seas or in deserts. One-third of the world's forestland disappeared in the same interval, and 40% of the total photosynthetic productivity is now co-opted by humans. Raven has noted that we humans seem to be treating the world as a business in the course of a liquidation sale.

The cost of this behavior will be wholesale species extinction. Raven estimates that in the last several decades, Ecuador has lost 2000 species of endemic plants to extinction. Extinction rates are booming. Raven, that night, gave the highest estimate of final extinction rate that I had ever heard: He estimates that 67% of all species now living on Earth will be extinct by the end of A.D. 2200—two centuries from now—unless radical steps are taken fast.

Raven was not simply transmitting a grim message without offering a solution. The actions he proposed near the end of his talk are both clear and difficult.

1. We *must* reduce human populations. Our planet cannot withstand 6 billion people without irreparable harm, let alone the 11 to 12 billion slated to live on this small planet by the middle of the next century.

2. We *must* preserve large tracts of land. Nature preserves where entire ecosystems are saved, not isolated attempts to save individual species, are the answer to preserving endangered species.

3. Humanity has to improve what might be called restoration ecology. One of the great challenges confronting us is to find a way to return logged areas to forest. This effort, however, will be useless unless we stabilize human population.

4. To feed the hungry human mouths and still retain large regions of undisturbed or restored wild regions, we will have to become far more efficient at agriculture. The great loss of soil cover even in the last 50 years is largely due to poor and wasteful agricultural practices—and especially to overgrazing. Equally damaging has been the rampant use of pesticides. It is clear that the wholesale use of DDT and many other pesticides is leading to widespread species extinction, as well as contributing significantly to the rise in human cancers. Raven suggests that only through breakthroughs in genetic engineering can we win the battle against crop-eating insects and

other pests. Manipulating the DNA of crops and their predators is the best way to achieve a pesticide-free world.

5. Finally, the imbalance in the distribution of scientific expertise must be redressed if these changes are to take place. "Developing" countries house 70% of the human population at the present time but are home to only 6% of the world's scientists. In more than 100 countries, there is *no* scientific or engineering establishment. We, the so-called developed countries, can recommend all the scientific solutions we want, but unless local scientists endorse these new solutions and put them into practice, no change will occur.

By the end of the evening I was silent, as were the other 200 in the audience. As the lecture ended, and the questions marking the end of such an event dwindled, I heard, from the adjoining lecture hall, the muffled laughs and applause as Oliver Sacks regaled *his* audience with tales of his inspiring patients. There was no such laughter coming from our hall—only the silence of troubled reflection.

. . . .

Raven had addressed one of the most pressing of all biological issues: biological diversity, a term often shortened to *biodiversity*. Were there more species in the past than now, or fewer? How rapidly do species numbers change? What controls the diversity of a given region? If we define diversity as the number of species present in any given area, can we arrive at some rough rule governing diversity? There are no simple answers to these questions, because many factors enter into the equations: nutrient availability, habitat type, and amount of water; factors that affect rates of speciation, such as rates of genetic change and the rates at which barriers form to isolate subpopulations of a species; and especially rates of extinction. Biologists have long recognized that diversity appears to be roughly related to habitat size, and this makes sense: The larger the area, the more animals and plants and, at the same time, the more *different kinds* of animals and plant that should be accommodated. But is extinction rate also related to habitat size? Do larger habitats, or only larger population sizes, protect individual species from extinction?

Some rough rules of thumb governing this relationship were first formulated by two famous ecologists, Robert MacArthur and E. O. Wilson, who in the 1960s proposed a new theory relating species diversity to habitat parameters. They called their proposal "the equilibrium theory of island biogeography." This theory related the area of habitat to the number of species present; as habitat area increases, species numbers also increase, and they do so in a predictable way. Similarly, as habitat area *decreases*, species numbers fall. Because the number of species bears a predictable relationship to the area available, deforestation leads to a shrinking of habitat in a way amenable to analysis. MacArthur and Wilson's equations can be used to predict rates of extinction. But MacArthur and Wilson's work showed an even more alarming result. In their studies on the number of species present on islands, they found that an island always has *fewer* species than a similarly sized mainland or continental area, even if the habitats are otherwise exactly identical. The implications of this are frightening. It means that parks and reserves, which essentially become islands surrounded by disturbed habitat, will always suffer a loss of species. It also means that cutting up the rain forest (or any forest) into patches of disturbed and undisturbed regions—creating many "islands" of forest—will greatly increase the rate of extinction.

These types of models are directly relevant to the issue of large-mammal extinction during the Pleistocene and to the plight of larger mammals of the present day. Significant direct relationships have been shown to have existed between area and diversity for a variety of taxa during many different intervals of geological time: The larger the habitat area, the more kinds of plants and animals that live within. And in every case where habitat area is reduced, the trends are the same. When habitat area diminishes, so does diversity. Reduction in diversity can take place by one of only two mechanisms: Either some creatures migrate to other habitats, or they go extinct. Either scenario leads to "local extinction," the de facto extinction of one or more species that previously lived in a given habitat.

With the changes in climate that occurred at the end of the Pleistocene, just such habitat fragmentation took place. The complex assemblage of vegetation that characterized much of the Ice Age began to fragment and disappear in many continents, most notably North America and northern Europe and Asia. More nutritious and diverse plant communities gave way to less diverse communities characterized by plants adapted to cooler, drier environmental conditions, and the remaining regions capable of supporting the grazing or browsing of the Ice Age megamammals decreased in size. With this reduction in habitat size, diversity began to drop.

One of the central predictions of the equilibrium theory of island biogeography is that a group of small refuges will preserve fewer species than a large refuge of equal total area. This tenet has been challenged on theoretical grounds. One point, however, is not in dispute: Any species needs some minimal habitat area, and when habitat size falls below this critical value, extinction occurs.

The actual causes of extinction, as populations reach some minimum size, thus may be related to the following five effects, acting singly or in concert.

1. For a great many demographic traits, such as the number of reproductively viable population members, variation is inversely proportional to population size. The smaller the population, the higher the risk that extinction will occur because of some chance event such as all population members being the same sex.

2. If a population is already small, then unusual values in physical factors (such as rainfall) or a change in biotic factors (such as a rapid increase in the number of predators) may be sufficient to cause extinction, whereas larger populations should weather such perturbations.

3. Natural catastrophes such forest fires, hurricanes, or greater temperature extremes are more likely to overwhelm smaller populations.

4. Dysfunction of social behavior can endanger small populations. If, for instance, large breeding aggregations are necessary for reproductive success, then some minimum number of individuals may be required.

5. Inbreeding depression may occur. First, the chance of the emergence of debilitating disease caused by some recessive gene increases as populations get smaller. Second, the gene pool itself becomes smaller in small populations, and inbreeding inevitably occurs. This can lead to loss of vigor and fertility, further diminishing population size.

Clearly, these factors will be of utmost importance in deciding whether a given game park is large enough to sustain various species, for a game park is nothing but a small island within a sea of humanity and human-perturbed ecosystems. We must plan now, with experience and knowledge gained not only from theoretical models but also from our understanding of the past. Perhaps in at least this one instance, as far as saving the elephants is concerned, the past is the key to the present.

. . . .

The plight and status of the African elephant have recently been succinctly analyzed by biologists Peter Armbruster and Russell Lande, both of the University of Oregon, who wrote in 1993,

> Overall numbers are in decline, and in several areas localized extinctions have already taken place or are imminent. In East Africa, few elephants are predicted to survive beyond 1995 outside high security areas, and a similar trend is predicted on the continent as a whole within 20 years. Habitat loss and poaching are the two factors most responsible for the African elephant's current decline, and both are a direct product of human population growth throughout Africa.

This passage could just as well be used to describe the factors that led to the extinction of the mastodon and mammoth so many thousands of years ago.

Armbruster and Lande used a slightly different type of kill curve in an attempt to determine how large a game reserve must be for a viable elephant population to be maintained. Their model also reflects a somewhat longer view than most conventional models. They ask the following question: "What is the reserve size necessary to preserve a population of elephants with a 99% probability that they will not go extinct in 1000 years?"

To find out, they employed a stochastic model that incorporates life history characteristics of the elephants (such as death rates and reproductive rates) as well as changing environmental conditions (such as rainfall and the lengths of droughts).

To assess the validity of the model, they first ran their computer simulation for a world where there is no environmental change. Eleven male and 11 female elephants were put into a park of 4000 square miles and "observed" for 625 years. The 22 original elephants grew to a population of over 11,000 elephants, and the population then remained constant at this number, or about 3 elephants for every square mile.

In their second run, the two scientists modeled the case where the environment is not constant; they imposed droughts or climatic fluctuations of varying intensity, but at values consistent with those observed in East Africa. In this case, instead of maintaining a constant number of elephants, the elephant population fluctuated wildly, peaking near 12,000 elephants and dropping to around 1000. Droughts cause rapid change in the success of the elephants. During times of greatest stress, only large initial populations survive, and then only at small numbers. Survival is thus related to habitat size. Armbruster and Lande concluded that only game parks of 1000 square miles or greater are large enough to ensure survival of the elephants.

There are magnificent game parks in East Africa. Twenty such large game parks are found in Kenya, Tanzania, Uganda, and Zimbabwe. Yet only 35% of these are of the minimum size that is apparently necessary to sustain African elephants for the next 1000 years. Smaller game parks could work if they were in areas of higher rainfall. However, areas of high rainfall have been co-opted for human agriculture.

Armbruster and Lande's model has one great simplifying feature: It assumes no human predation on elephants. Sadly, elephants have made one colossal evolutionary mistake. They have evolved a substance coveted by humans, ivory.

The harvesting of tusk for ivory led to a precipitous decline in elephant numbers. It is estimated that from 1973 to 1980 alone, two-thirds of all elephants in Kenya were killed for ivory. In 1989 the Convention of International Trade in Endangered species (CITES) listed the Africa elephant as an endangered species, thus banning trade in ivory. Nevertheless, poaching continues because of the immense value of the tusks compared to the very low per-capita income in most regions native to elephants. Human predation thus becomes an additional factor affecting elephant populations.

Here again we find striking parallels between the fates of the ancient mammoth and mastodon and the modern-day elephant. With the end of the Ice Age, the ancient elephants lived in increasingly fragmented habitats, just as today's elephants do. Within these islands of favorable habitat, the declining elephant populations are especially vulnerable to human predation, even if that predation seems to occur at a very low rate.

If we consign elephants to parks and reserves, those parks must be large enough to withstand the rare droughts and environmental changes that are surely to come. They must be secure from human predation as well. Perhaps we could best sustain the elephants by genetically altering them, making them tuskless and distasteful to our palate.

It is in the next century that the danger looms large. There are 5.5 billion people on Earth today, and the world will have to support about twice that many within a century. Humans and great elephants have long coexisted. But will we last—together—another millennium, let alone another 100,000 years? I feel confident that humans will make it. About the elephants, I am not so sure.

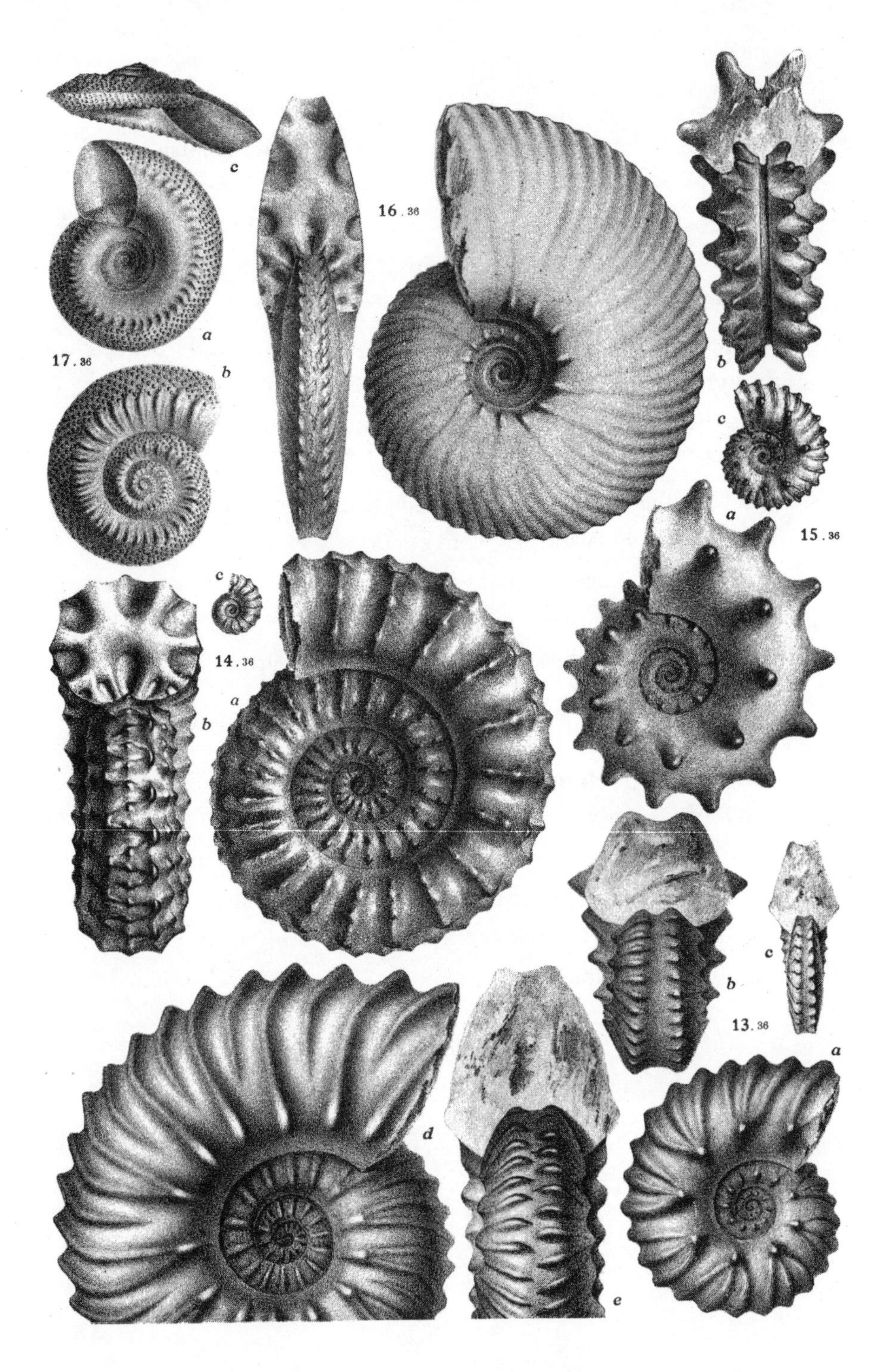

Ammonite shells from Europe. All of these died out with the dinosaurs, and illustrate the reality of extinction.

10

The Smoking Gun

It is strange how certain phrases stick with you. When I was a boy, my parents spent much of their leisure time reading murder mysteries—"whodunits," as my mother called them. It was an American mania, and the books seemed to be written by formula. Victim dies, wrong suspect charged, detective brought in to solve case. And, of course, there was always a "smoking gun" that ultimately implicated the true murderer.

Later, the phrase moved out the genre of murder mysteries and was applied in other contexts, including the pursuit of Richard Nixon during the Watergate trials. The phrase even entered the long controversy over the dinosaurs' extinction. Soon after the announcement by the Alvarez group that a meteor had killed the dinosaurs, critics brought up the lack of a "smoking gun," saying that until a crater was found, the theory would remain unproved.

The same argument is now used with regard to the Ice Age extinctions. There is no smoking gun, critics of Paul Martin's overkill hypothesis proclaim. There is no dramatic evidence to prove conclusively that humans (or climate change, or anything else, for that matter) really did it. There are no bone beds, no fields of mammoth corpses riddled with Clovis spear points.

Unfortunately, there is rarely a smoking gun in science, especially when one is dealing with large and extraordinarily complex events such as mass extinctions, which took thousands of years or more to unfold. Yet in the case of the dead mammoths and mastodons, a recent discovery comes as close as we will probably ever get to the smoking gun that can pin guilt on the murderer of these Ice Age victims. It comes from one of the most clever and astute of all paleontological detectives: a charming and disarming man named Dan Fisher, who I believe has finally brought this 10-millennium-long murder mystery to a close.

I have written about Dan Fisher before. Now at the University of Michigan, he has long been a specialist in fossils of animals without backbones. He spent his early career studying horseshoe crabs, those odd, crustacean-like living fossils found along the eastern seaboard of North America. But like most paleontologists, Dan has wide interests, and fossils emerging from the local glacial deposits near his Michigan home have caught his attention. Dan, like so many others, has heard and responded to the siren call of distant mammoths.

Mammoth bones are common in the glacial tills of Michigan. Like the Pacific northwest, most of Michigan was repeatedly covered by glacial ice during the Ice Age. When that ice retreated for the last time some 15,000 years ago, it left behind a world inhabited by giant mammals, including mammoths and mastodons, and eventually humans. Fossil skeletons of that period now emerge from eroding bogs, gravel deposits, and even golf courses. Sometimes those skeletons show evidence of having been butchered. Humans and extinct elephant species clearly inhabited the same regions of the American midwest around 11,000 years ago, and they knew each other well.

Dan Fisher has been studying the remains of the mastodons and mammoths discovered in Michigan and neighboring Ohio for well over a decade. Like many other scientists, he has come to the conclusion that the fossil skeletons found in association with human artifacts are evidence of a behavioral association of some kind between humans and the now-extinct proboscideans. But was it the association of hunter and hunted or of scavenger and the already dead? As the anthropologist Gary Haynes put it, did the mammoths and mastodons of Michigan die off, or were they killed off? Fisher, along with so many of us seeking the answer to that question, has been frustrated by the limited information available in any single pile of fossilized bones emerging from the ancient Ice Age gravels. But Fisher is also one of the smartest humans on this planet, and he has devised an entirely new way of examining the problem. From within the bones and tusks of these fossils, he has coaxed volumes of previously unrecovered information that tells new tales about the lives of humans and ancient elephants at the end of the Ice Age.

Fisher has concentrated on the fossil ivory tusks of the mammoths and mastodons. These huge, curving teeth grow in a manner somewhat analogous to trees; they have growth rings that are secreted annually. These rings can tell us much about the life of the animal, just as tree rings offer information about the life of a tree. Tree rings and proboscidean tusk rings are studied in the same way: The tree (or tusk) is cut perpendicular to its length, and the cut face is polished. The spacing between the growth rings can then be measured.

During times of environmental stress and drought, tree rings are close together because little growth has taken place. During good times, however, when the tree is undergoing rapid growth as a result of an abundance of water, ample nutrients, and favorable temperatures, the rings are widely spaced. The tusks of elephants, mastodons, and mammoths respond in similar fashion. During periods of high growth, they are well spaced; during times of food deprivation, they are cramped together. And elephant tusks reveal far more about the life of the elephant than a simple chronology of good times and bad. Elephant tusk rings, and the chemicals locked within them, can offer accurate clues not only to how much an individual ate but also what it ate. The types of food available to the creatures are revealed by the isotopic fractions of nitrogen and carbon locked up in the tusks, for these isotopes are found in differing percentages in various plants eaten by the proboscideans.

For a female elephant, birth events are also recorded in the tusks. The tusk rings reveal how many times in her life she reproduced and the length of time between births. During pregnancy, a female elephant grows little tusk, because the calcium normally used in tusk formation is diverted to bone formation in the fetus. Tusk growth resumes once the baby is born. Thus pregnancy produces a readily observable gap in the tusks' growth record.

Fisher has called the tusks a diary of the animal's life. Reading enough of these diaries has enabled him to piece together a picture of conditions existing in the Michigan of more than 10,000 years ago. That picture yields new and valuable clues to the identity of the ancient murderer.

As we have seen throughout this book, there are but two competing theories to explain the extinction of the mammoths and mastodons: climate change and human hunting. The climate change hypothesis does not suggest that actual weather changes killed off the great mammals but that the changeover from a glacial to a warmer climate stimulated the rise of forest and the loss of the grasslands necessary for the diet of the mammoths and mastodons. Under this scenario, the extinction was caused by starvation, as the last bands of mammoths and mastodons searched for ever-smaller patches of favorable food supplies and died off in the process. Yet in October of 1996, at the annual gathering of the Society of Vertebrate Paleontologists, held in New York City, Fisher reported that the last mammoths and mastodons were apparently fat, fit, and well fed. The tusks examined show no evidence that any of them died of starvation. At least in the areas he has studied, Fisher has effectively demonstrated the falsity of one prediction of the climate change hypothesis. The mammoths and mastodons did not die off by starvation.

But were they killed off? Here too the tusks have yielded new and fascinating information. Like Gary Haynes, Dan Fisher has studied the behavior of elephants in Africa as a means of interpreting the way of life of ancient mammoths and mastodons. Modern elephants potentially face each of the supposed killers from the end of the Ice Age. The first of these is drought, which leads to starvation. Many elephant herds do encounter climate change, and of the various hardships that climate change inflicts, loss of food during dry conditions is the greatest killer. Yet modern elephants encounter a second killer as well: They are also threatened by human hunting.

Under each of these environmental challenges—starvation and predation—elephant herds respond reproductively in quite different ways. Under starvation conditions, elephants reproduce infrequently. It makes no evolutionary sense to bring young elephants into a world where adults are starving, and hence the elephants breed less frequently during long periods of drought and low food supplies. The young that are produced, if they survive, grow slowly. In contrast, when herds are being hunted with regularity, they reproduce more often and replenish their losses. The young that are born tend to grow and mature quickly; thus they too can breed as early as possible.

The tusks from 11,000-year-old female mammoths and mastodons of Michigan tell a grim tale. The females were well fed and so were probably under no stress from lack of food. But they showed very frequent breeding. Births were occurring every 4 years in the extinct forms, which is precisely the rate we see in modern elephants subject to constant hunting. Dan Fisher's work strongly suggests that the last of the mammoths and mastodons were "trying" to make up their numbers as rapidly as possible. Lack of food was not their problem. They were not dying off. They were being killed off. Perhaps some carnivore other than humans was the predator. But if so, these predators were invisible or boneless, for by the time of the last mastodons and mammoths in North America, the great saber-toothed tigers and cave bears, perhaps the only carnivores large enough to kill a mammoth, were already long extinct.

It is not hard to envision the scenario of that long-ago time. As the mammoths and mastodons became ever fewer in number, the humans who depended on them for food intensified their searching and hunting. And it is likely that ever more people continued to pour in from the northern regions, having heard tales of the warm lands to the south where the great mammoths and mastodons still lived.

There is very little "proof" in science. We have not yet found a *Triceratops* dinosaur fossil with one of its horns knocked off by the incoming asteroid that killed off the dinosaurs. Nor have we found an issue of the *Bedrock Daily News* from 10,000 years ago that features an article telling how the last mammoth of Michigan had been killed off by a local hunting party. There will always be some doubt. But in my mind, in Dan Fisher's mind and in many other scientists' minds as well, this particular scientific murder mystery is solved. We know "whodunit." We did it. Our species, our kind—humanity—armed only with stone-tipped spears, caused the extinction of the great mammoths and mastodons and perhaps that of many other large megamammal species. We did it simply by killing off about 2% of the population per year, year after year. Extinction debts are bad debts, and when they are eventually paid, the world is a poorer place.

11

Afterword: 3001

TIME IS NOW SHORT for me. That seems like a contradiction, for with my time machine I should have all the time in the world, and all of the many past worlds' time as well. But even the imaginary time traveler has constraints of life, and work, and love. There is now only one more trip to take. Like the hero of H. G. Wells's tale, or better yet like Charles Dickens's time traveler, I need to move forward in time, to see the Ghost of a far future Christmas. For my last trip I need to see the future: our future, and that of the elephants.

Silly, of course. The time machine I pilot, composed of my rock pick and paleontological tools, is useful only in the past. But I can imagine and synthesize. The future is one of multiple possibilities, some dim, some bright. Perhaps our planet will be hit again by a chance comet in the not-so-distant future, causing massive destruction and extinction, rewriting all the rules in the future game of evolution. If Earth is spared such a cataclysm, perhaps the future will be bright. Perhaps I am wrong, and evolution will create a more intelligent and wise branch of our species, a new *Homo sapiens* dedicated to thoughtful stewardship of Earth and its creatures. Or perhaps no such change will occur, and it is the current stock of our species that will weather the coming storm of human overpopulation and somehow negotiate the sea of storms in a warming world, with the riotous atmosphere that a worldwide temperature change would surely invoke. Perhaps.

Or perhaps H. G. Wells was right, We will become the Eloi, those timid creatures of Earth's surface wholly dependent on their supposed servants, the monstrous Morlocks of the subterranean regions. But *our* future Morlocks are not a new human subspecies but the environmental and ecological consequences of the 20th century, a time bomb reverberating far into the future. Perhaps the Morlocks will be the greenhouse gases we are now releasing into the atmosphere, the plague of human overpopulation whose seeds are even now being sown, the industrial waste and pollution we currently pour into the seas, the felling of the world's great forests, and the path to extinction faced by a majority of species currently living on the planet. Perhaps the Morlocks are those humans who refuse to see the storm warnings. I remember sending the outline of my last book to an editor who soon told me that although she loved the idea, she couldn't *possibly* publish a book about an impending mass extinction unless I gave it a happy ending. No one will buy a book unless it has a happy ending, she repeated over and over. Hail the Morlocks, and fear them.

So I will set my machine for one last journey, one last trip: this time 1000 years into the future. I will travel to Amboseli National Park in Kenya, site of one of the largest herds of elephants living in my world, to see what there is to see.

I arrive and encounter . . . order: rectangular farms as far as the eye can see and neatly dressed children playing and laughing among the geometrical rows of crops. The plants are like nothing I have ever seen, for there appear to be vegetables growing out of cactus plants. And the neat rows of wheat look anomalous in this African countryside. Everywhere about me there is an alien quality to the plants; they are genetically engineered food growing from drought-resistant root stock. There is no livestock in sight: humankind seem to have become vegetarians.

I walk toward a village visible in the distance, wondering where the park and its wildlife have gone. Adults as well as children move about the village, and I see many older people, a rare sight in the Africa of my time. All are well clothed, well fed, clean and apparently happy. I see no evidence of malnutrition; no gaunt skeletons in the last stage of AIDS, or slims disease, as they called it here in the 1990s; no evidence of malaria, the omnipresent staph infections, dengue fever, or any of the other plagues that afflicted Africa in the late 20th century. The huts are modern and clean; they have running water. There are solar power cells on each roof, but no cars, only large bicycle-like creations. The people look at me with curiosity but are too polite to stare, and I finally make my way to a schoolhouse in the village. The schoolmaster meets me gravely and asks me to tea. I cannot help but notice that he looks exactly like a hunter I encountered on my first time-traveling trip, back to the dawn of our species's existence. Outwardly we have not changed much in the last 100,000 years. But there are many more of us.

We sit in the shade, for the day is hot here in equatorial Africa. I tell him that I am from the past, from 1000 years ago, and he nods sagely, humoring the white-skinned lunatic before him. He is surprised at my skin color, though, for he has not seen anyone as light-skinned as I in his lifetime. Skin cancer killed off most of the white people ages ago, he says. I ask him about the history of the last thousand years. "Where is the park?" I ask. "Park?" he replies. "Oh, that has been gone for a very long time. Legend says that giant animals once roamed this land, but as you see they are now gone, long gone. We find their bones in our fields, once in a while, when plowing. But there are no wild beasts around here, thank goodness. They would eat our crops."

My host gives me a rough idea about the events of the 21st century, when human population swelled to over 12 billion. The parks were soon overrun by starving humans and by farmers needing land; the game was eaten. Elsewhere, the last rain forests were cut for farmland, after the great grain regions of the United States and central Asia became deserts during the climatic changes brought about by global warming. In the 22nd century, mass starvation and disease caused the population at last to drop; it was also a time of rising sea level, which drowned many of the productive deltas and rich farmlands around the world. Fresh, unpolluted water replaced oil as the most valuable liquid on earth. Genetic engineering created new plants, but with the loss of so much prime farmland, the loss of freshwater lakes to acidification, and the pollution of many aquifers, nearly all arable ground with access to untained fresh water had to be cultivated. The last wild places vanished beneath the plow. But peace came to the earth. Nations no longer competed with each other or rattled their sabers. The golden age of prosperity had finally arrived—at least for humans.

For a long time, the zoos of the world tried to maintain breeding stocks of the displaced wild animals. With great effort, they kept populations viable for a century or so, in the hope that *someday* the beasts of the earth could be reintroduced into the wild. But eventually it became clear that there was no wild to put them into, and people had other preoccupations. It got too expensive, and the remaining specimens were finally stuffed. The wild animals and the wild regions had vanished in this age of humanity.

I needed to see other areas, other countries, to find out how the rest of the world had fared. "How can I travel?" I asked. "I need to wander." My host looked at me with alarm. "Wander? Why would you do that? It is the same everywhere. There is no need to wander."

I had one final question, though I feared the answer. "Where are the elephants?" I asked. "Elephants?" my new friend replied. "Why, they're extinct, of course. We have legends of them and what they did to our crops. But didn't they die out with the dinosaurs?"

References

Alvarez L, Alvarez W, Asaro F, and Michel H. Extra-terrestial cause for the Cretaceous-Tertiary extinction. Science 1980;208:1094–1108.

Antevs E. Artifacts with mammoth remains, Naco, Arizona. American Antiquity 1953;19:15–8.

Armbruster P and Lande R. A population viability analysis for African elephant (Loxodonta africana): how big should reserves be? Conserv Biol 1993;7:602–10.

Axelrod D. Quaternary extinctions of large mammals. Univ Calif Pubs Geol Sci 1967;1–42.

Beck M. On discerning the cause of late Pleistocene megafaunal extinctions. Paleobiology 1996;22:91–103.

Bourgeois J. Tsunami deposits and the K/T boundary. A sedimentologist's perspective. Lunar Planetary Institute Contribution 1994;825:16.

Burenhult G, ed. The First Humans: Human Origins and History to 10,000 BC. NY: Harper Collins, 1993.

Covey C, Thompson S, Weismann P, and Maccracken M. Global climatic effects of atmospheric dust from an asteroid or comet impact on earth. Global Planet Change 1994;9:263–73.

Dobson A. Conservation and biodiversity. NY: Scientific American Library, 1996.

Donovan S, ed. Mass Extinctions Processes and Evidence. NY: Columbia University Press, 1989.

Eisley L. Archaeological observations on the problem of post-glacial extinction. American Antiquity 1938;8:209–17.

Eldredge N and Gould S. Punctuated equilibrium: an alternative to phyletic gradualism. In: T Schopf, Models in Paleobiology. NY: WH Freeman, 1972.

Ellis J and Schramm D. Could a supernova explosion have caused a mass extinction? Proc Natl Acad Sci 1995;92:235–8.

Erwin D. The Great Paleozoic Crisis: Life and Death in the Permian. NY: Columbia University Press, 1993.

Fagan B. The Great Journey: The Peopling of Ancient America. NY: Thames and Hudson, 1987.

Gould S, ed. The Book of Life. NY: WW Norton, 1993.

Graham R. Diversity and community structure of the late Pleistocene mammal fauna of North America. Acta Zool Fenn 1985;170:181–92.

Grayson D. Archeological associations with extinct Pleistocene mammals in North America. J Archeol Sci 1984;11:213–21.

Grayson D. Late Pleistocene mammalian extinctions in North America: taxonomy, chronology, and explanations. J World Prehistory 1991;5: 193–231.

Greenland Ice Corps Project. Climate instability during the last interglacial period recorded in the GRIP ice core. Nature 1993;364:203–7.

Haury E. Artifacts with mammoth remains, Naco, Arizona. American Antiquity 1953;19:1–14.

Haynes CV. Fluted projectile points: their age and dispersion. Science 1964;145:1408–13.

Haynes G. Mammoths, Mastodons, and Elephants. NY: Cambridge University Press, 1995.

Kingdon J. Self–Made Man: Human Evolution from Eden to Extinction. NY: John Wiley, 1993.

Jablonski D. Extinctions in the fossil record. In: J Lawton and R May, eds. Estimating Extinction Rates. NY: Oxford University Press, 1996.

Johanson D and Edey M. Lucy: The Beginnings of Humankind. NY: Morrow, 1981.

Jennings F. The Founders of America. NY: WW Norton, 1993.

Keller G. Global biotic effects of the K/T boundary event: mass extinction restricted to low latitudes? Lunar and Planetary Institute Contribution 1994;825:57–8.

Leakey R and Lewin R. Origins. NY: Doubleday, 1977.

Leakey R and Lewin R. Origins Reconsidered: In Search of What Makes Us Human. NY: Doubleday, 1992.

Leakey R and Lewin R. The Sixth Extinction: Patterns of Life and the Future of Humankind. NY: Doubleday, 1995.

Lister A and Bahn P. Mammoths. NY: Macmillan Co., 1994.

Martin P. The Last 10,000 Years: A Fossil Pollen Record of the American Southwest. Tuczon: University of Arizona Press, 1963.

Martin P and Wright H. Pleistocene Extinctions: The Search for a Cause. New Haven, CT: Yale University Press, 1989.

Martin P. The Discovery of America. Science 1973;179:969–74.

Martin P and Klein R, eds. Quaternary Extinctions. Tuczon: University of Arizona Press, 1989.

Marshall C. Confidence Intervals on Stratigraphic Ranges. Paleobiology 1990;16:1–10.

Marshall C and Ward P. Sudden and gradual molluscan extinctions in the latest Cretaceous of Western Europe Tethys. Science 1996;274:1360–3.

May R and Nowak M. Superinfection, metapopulatioin dynamics, and the evolution of diversity. J Theory Biol 1994;170:95–114.

Pielou E. After the Ice Age. Chicago: The University of Chicago Press, 1991.

Rampino M and Caldeira K. Major episodes of geologic change: correlations, time structure, and possible causes. Earth Planet Sci 1993;114:215–27.

Raup D. A kill curve for Phanerozoic marine species. Paleobiology 1991;17:37–48.

Raup D. Extinction: Bad Genes or Bad Luck? NY: WW Norton, 1991.

Retallack G. Permian–Triassic crisis on land. Science 1995;267:77–80.

Rudwick M. The Meaning of Fossils. Canton, MA: Neale Watson Publishing, 1995.

Sharpton V, Martin L, and Schuraytz B. The Chicxulub multiring basin: evaluation of geophysical data, well logs, and drill core samples. Lunar Planetary Institute Contribution 1994;825:108–9.

Sheehan P, Fastovsky D, Hoffman G, Berghaus C, and Gabriel D. Sudden extinction of the dinosaurs: latest Cretaceous, Upper Great Plains, U.S.A. Science 1991;254:835–9.

Sigurdsson H, D'hondt S, and Carey S. The impact of the Cretaceous-Tertiary Bolide on evaporite terrain and generation of major sulfuric acid aerosol. Earth Planetary Sci Lett 1992;109:543–59.

Stanley S. Extinction. NY: Scientific American Books, 1987 .

Stanley S. Earth and Life Through Time. NY: WH Freeman, 1989.

Walker A and Leakey R. The hominids of East Turkana. NY: Scientific American, 1978:239.

Ward P. The Cretaceous/Tertiary extinctions in the marine realm: a 1990 perspective. Geol Soc Am (Special Paper) 1990;247:425–32.

Ward P. On Methuselah US Trial: Living Fossils and the Great Extinctions. NY: WH Freeman, 1991.

Ward P. The End of Evolution. NY: Bantam Books, 1995.

Wilson E. The Diversity of Life. Boston: Harvard University Press, 1992.

Index

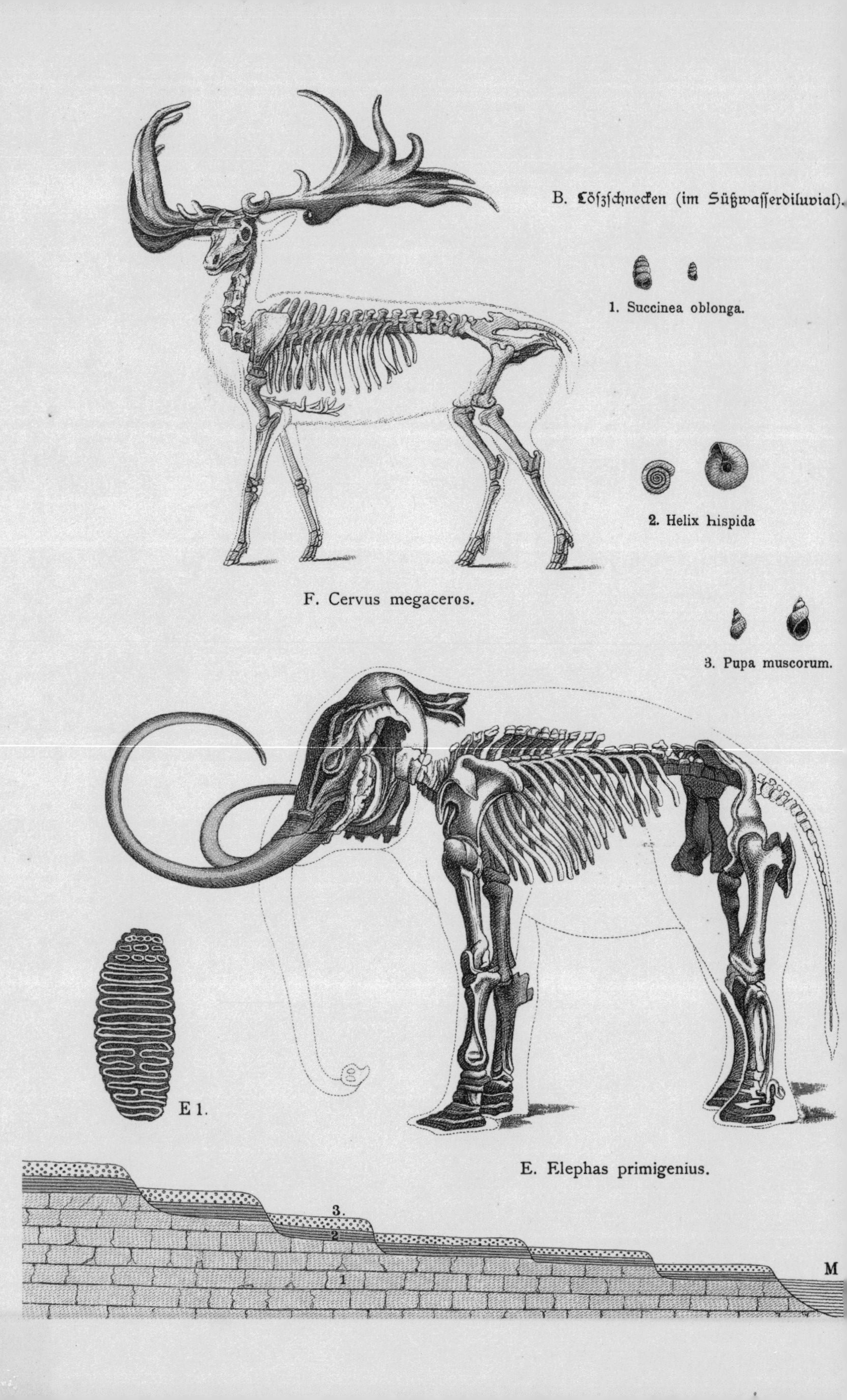
B. Lößschnecken (im Süßwasserdiluvial).
1. Succinea oblonga.
2. Helix hispida
3. Pupa muscorum.
F. Cervus megaceros.
E 1.
E. Elephas primigenius.
3.
2
1
M